Automatic Flight Control Systems - Latest Developments

Editor

Beatrice Adamsen

scitus
academics

Automatic Flight Control Systems - Latest Developments
Edited by **Beatrice Adamsen**

ISBN: 978-1-68117-273-6
Library of Congress Control Number: 2016937444

SCITUS Academics LLC,
www.scitusacademics.com
Box No. 4766, 616 Corporate Way,
Suite 2, Valley Cottage,
NY 10989

Printed in United States of America on Acid Free Paper

Preface

Flight control history is inseparably linked to aviation history itself. Since the early years, the concept of automatic flight control systems has evolved from mechanical control systems to highly advanced automatic fly-by-wire flight control systems currently available in military jets and civil airlines. A conventional flight control system for fixed aircraft consists of flight control surfaces, the respective cockpit controls, connecting links and the operating mechanisms necessary to control the direction of the aircraft during flight.
The controls of the aircraft engine are also considered flight controls as they change speed. An autopilot is a system used to control a vehicle's trajectory without constant "hands-on" control by a human operator. Autopilots do not replace a human operator but help them to control the vehicle, allowing them to concentrate on broader operational aspects, such as trajectory monitoring, weather and system. Autopilots are used for aircraft, spacecraft, missiles and other applications. Autopilots have evolved significantly over time, from early autopilots that only had an attitude towards modern autopilots capable of automated landings under the pilot's supervision. In a modern large aircraft, the autopilot typically reads the position and attitude of the aircraft from an inertial control system.

This book emphasizes a selection of important research areas, such as inertial navigation, aeroservoelastic control, fault tolerant flight control, adaptive flight control, and control of unmanned aircraft and helicopters, unmanned space re-entry vehicle trajectory control, aeroservoelastic control, and adaptive flight control.

TABLE OF CONTENTS

CHAPTER 1

AUTOMATIC FLIGHT CONTROL SYSTEM DESIGN OF LEVEL CHANGE MODE FOR A LARGE AIRCRAFT

Huajun Gong[1], Ziyang Zhen[1*], Xin Li[1], Ju Jiang[1] and Xinhua Wang[1]

College of Automation Engineering, Nanjing University of Aeronautics and Astronautics, China

ABSTRACT

The level change mode is an essential part of large civil aircraft automatic flight control systems. In cruise, with the decrease of the plane's weight caused by fuel consumption and the influence of bad weather, such as thunderstorms, the level change mode is required to solve this problem. This work establishes a nonlinear model of large aircraft, takes level changed from 9500m to 10100m as an example to design control laws for the level change mode in cruise. The classical engineering method is used to design longitudinal and lateral control laws synthetically. The flight qualities are considered in the design process. Simulation results indicate the control laws can meet design requirements and have a good anti-gust performance.

KEYWORDS

Large Aircraft, Level Change, Attitude Control, Classical Control

1. INTRODUCTION

The fly-by-wire automatic flight control system has obvious advantages such as improving handling characteristics, enhancing flight safety, reducing the pilot's operating burden, increasing economic efficiency and so on, so it is an essential part of a large civil aircraft. The automatic flight control system produces and transmits automatic control instructions to achieve automatic or semi-automatic flight path control, or to control the aircraft to restrain all kinds of disturbances automatically. Autopilot, the driving rod/reel manipulation, automatic throttle and structural mode control all constitute the automatic flight control system. In modern aviation, civil aircraft from all over the world all have many kinds of control modes, and can achieve many kinds of flight tasks. So in the phase of automatic flight control system designing, for each mode, it needs a flight control engineer to design control laws that meet the requirements of performance index, which effects the quality of the designed control laws and directly impacts flying qualities and flight safety. In an automatic flight control system, the mode selection panel (MSP) produces a control mode choosing the interface for the pilot. Generally, all control modes can be divided into three categories: vertical mode, horizontal mode and automatic throttle mode. The level change (FLCH or LVL CHG) mode studied in this paper belongs to the vertical mode.

In actual flight, the aircraft is required to change its level according to its own weight, weather conditions and other factors. Generally, the larger the aircraft's weight, the lower the flight level. So with fuel consumption, the aircraft needs to increase cruise level at the same time, and to avoid bad weather such as thunderstorms and upper air conditions, it also needs to change level. The General Flight Rules of the People's Republic of China put in force from November 22, 2007, stipulate that when true course angle is in the range from 0 degrees to 179 degrees, and when height is from 900 m to 8100 m, every 600 m is a flight level. When the height is from 8900 m to 12500 m, every 600 m is a flight level. When the height is above 12500 m, every 1200 m is a flight level. On the other hand, when true course angle is in the range from 180 degrees to 359 degrees, and when the height is from 600 m to 8400 m, every 600 m is a flight level. When the height is from 9200 m to 12200 m, every 600 m is a flight level. When the height is above 13100 m, every 1200 m is a flight level. The flight level equipped standard diagram is shown in Figure 1.

There are not enough studies on both design longitude and lateral control laws, or testing of the control systems' anti-disturbance abilities. Reference [1] uses nonlinear energy-based control method to achieve civil aircraft automatic landing, where the control system is able to cope with various flight conditions and the stability and performance robustness is guaranteed. Reference [2] applies linear parameter-varying (LPV) control synthesis to design fault tolerant controllers for Boeing 747-100/200. Reference [3] uses the adaptive sliding-mode control method to design the flight path tracking control law of civil aircraft. This method can improve flight performance and at the same time restrain disturbance and model uncertainty. Reference [4] uses the fuzzy control method to design an automatic landing control law. Simulation results indicate the designed control law can achieve flexible and effective automatic landing, and the control system has a better anti-disturbance performance. Reference [5] uses the μμ synthesis method to design a robust flight control law. This control system has better dynamic performance and robustness, but the anti-gust ability has still not been tested. Reference [6] uses the LQG/LTR method to design a longitudinal control augmentation system (CAS), which enhances stability, robustness and anti-disturbance capacity, but the simulation is also based on a linear system.

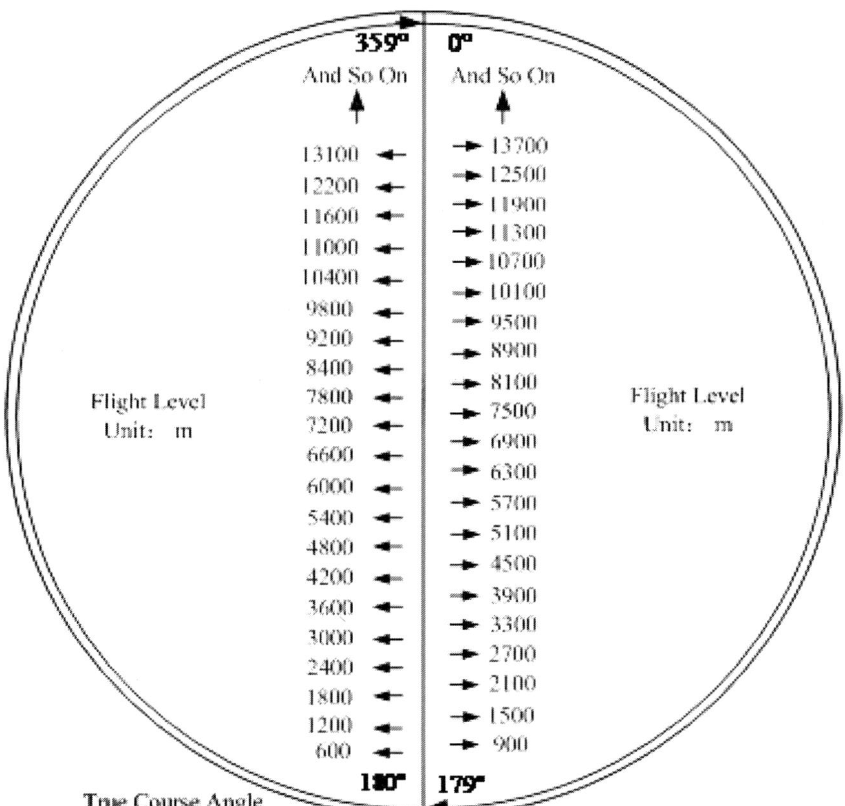

Figure 1. Flight level equipped standard diagram

This study focuses on the design of an automatic control system for the level change mode of large aircraft, by using a classical engineering method. This paper is organized as follows. In section 2, a nonlinear model with total variables is obtained for controller design and simulation tests, the linear model is derived and the natural properties are investigated. Section 3 designs a trajectory control system for level change mode and the relative controllers. A simulation study is implemented in section 4. Finally, the conclusions and the acknowledgements are given in section 5 and section 6.

2. MATHEMATIC MODEL OF LARGE AIRCRAFT

2.1. Nonlinear Model

Based on the aerodynamic parameters of the Boeing707, a nonlinear model with total variables is established. The structure of the nonlinear model of the large aircraft is shown in Figure 2 and the variables' definitions are shown in Table 1.

2.2. Linear Model

Level change mode control laws for large civil aircraft in cruise phase are studied in this work. The simulation is in condition of flight level changing from 9500m to 10100m, during the climbing phase. When height changes, so too do air density and air temperature, therefore model parameters will change. However, the height only has small change, thus the model parameters change not too large. So parameters of the middle section can be used to establish the mathematic model of the whole process. The selected trim point is that the cruise height equals 10000m, air speed equals

240m/s, pitch angle and angle of attack both equal 0.26°, and at this time, longitudinal linear formula is shown as

$$\dot{x} = A_{lon}x + B_{lon}u \tag{1}$$

here, state vector $x = \begin{bmatrix} \Delta V & \Delta\alpha & \Delta q & \Delta\theta \end{bmatrix}^T$, input vector $u = \begin{bmatrix} \Delta\delta_e & \Delta\delta_T \end{bmatrix}^T$, unit of ΔV is m/s, units of Δα, Δθ and Δδe are all rad, unit of Δq is rad/s, ΔδT is opening degree of throttle, its unit is percentage form and the coefficient matrices are as follows.

$$A_{lon} = \begin{bmatrix} -0.0050 & 6.5164 & 0 & -9.800 \\ -0.0004 & -0.6130 & 1.0000 & -0.0000 \\ -0.0008 & -2.4916 & -0.2822 & -0.0000 \\ 0 & 0 & 1 & 0 \end{bmatrix}$$

$$B_{lon} = \begin{bmatrix} 0.3504 & 6.4159 \\ -0.0251 & -0.0001 \\ -1.8921 & -0.1418 \\ 0 & 0 \end{bmatrix}$$

$$\dot{x} = A_{lat}x + B_{lat}u$$

The linearization formula of the lateral movement is shown as

$$\dot{x} = A_{lat}x + B_{lat}u \tag{2}$$

Table 1. Variables' notation

Variables	Notation	Variables	Notation
X_b	Axial force	α	Angle of attack
Y_b	Sideforce	β	Sideslip angle
Z_b	Normal lift force	θ	Pitch attitude
V	Airspeed	ϕ	Roll attitude
L_b	Rolling moment	δ_e	Elevator angle
M_b	Pitching moment	δ_T	Throttle opening
N_b	Yawing moment	δ_a	Aileron angle
p	Roll rate	δ_r	Rudder angle
q	Pitch rate	r	Yaw rate

here, state vector $x=[\beta \quad p \quad r \quad \phi]'$, input vector $u=[\delta_a \quad \delta_r]'$, the units are the same as that in formula (1) and the coefficient matrices are as follows.

$$A_{lat}=\begin{bmatrix} -0.1027 & 0.0015 & -0.9938 & 0.0408 \\ -6.3489 & -2.5099 & 1.8116 & 0.0000 \\ 2.5033 & -0.1917 & -0.4088 & 0.0000 \\ 0.0000 & 1.0000 & 0.0045 & 0.0000 \end{bmatrix}$$

$$B_{lat}=\begin{bmatrix} 0.0000 & 0.0324 \\ -3.0233 & 0.8795 \\ 0.0000 & -2.3229 \\ 0.0000 & 0.0000 \end{bmatrix}$$

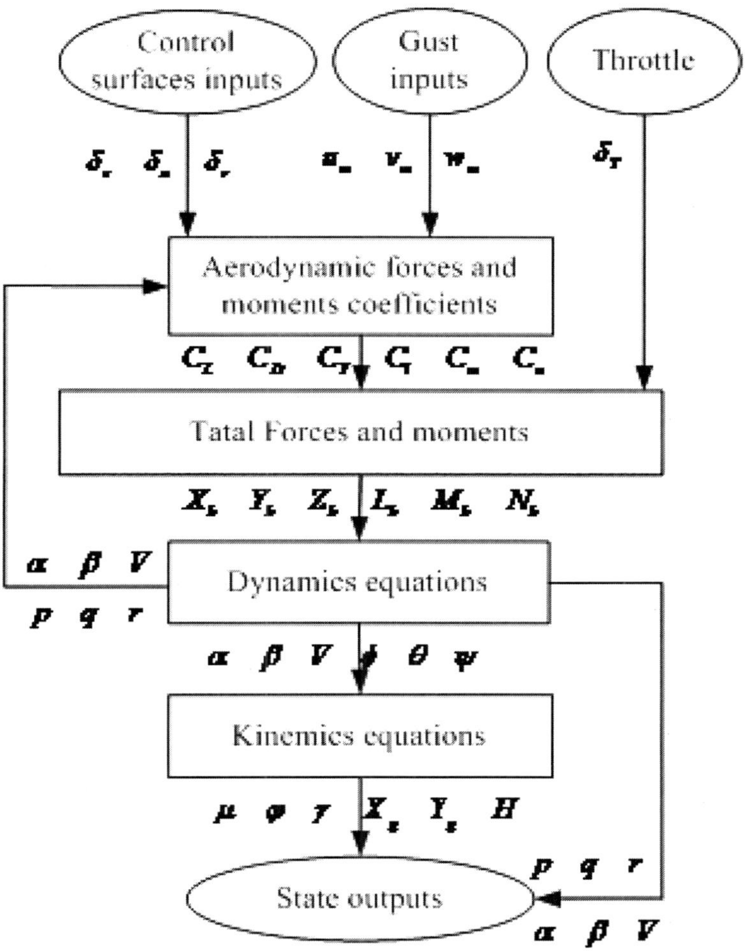

Figure 2. Nonlinear model structure of large aircraft

Considering the actual conditions, the transfer functions of engine and throttle servo are respectively given by

$$G_E = \frac{\Delta\delta_T(s)}{\Delta u_{\delta_T}(s)} = \frac{1}{s+1} \tag{3}$$

$$G_S = \frac{\Delta u_{\delta_T}(s)}{\Delta u(s)} = \frac{1}{0.05s+1} = \frac{20}{s+20} \tag{4}$$

The transfer functions of elevator servo system, aileron servo system and rudder servo system are shown as (5). Both of them are inertial first-order systems whose time constants are 0.1s.

$$G_{\delta_e} = G_{\delta_a} = G_{\delta_r} = \frac{10}{s+10} \tag{5}$$

The main task of this paper is designing a control law of level change mode for the Boeing 707 in the cruise phase. At first, the simplified transfer function should be obtained by the linear formulas. The simplified transfer functions of longitudinal short-period motion are shown in (6) and (7). The simplified transfer function of longitudinal long-period motion is shown in (8).

$$G_{\delta_e}^{\theta} = \frac{\Delta\theta(s)}{\Delta\delta_e(s)} = \frac{-1.8921s-1.0976}{s\left(s^2+0.8952s+2.6646\right)} \tag{6}$$

$$G_{\delta_e}^{\alpha} = \frac{\Delta\alpha(s)}{\Delta\delta_e(s)} = \frac{-0.0250s-1.8992}{s^2+0.8952s+2.6646} \tag{7}$$

$$G_{\delta_T}^{V} = \frac{\Delta V(s)}{\Delta \delta_T(s)} = \frac{6.4159s+3.9323}{s^2+0.6180s+0.0057} \quad (8)$$

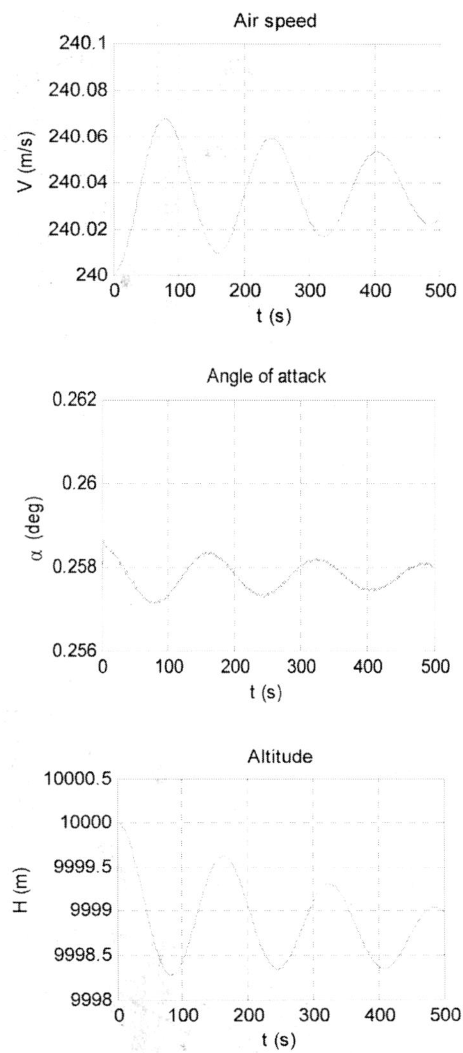

Figure 3. Flight state responses under trim conditions

2.3. Natural Properties Analysis

Give the trim values of the large aircraft, the elevator angle is -1.06deg, throttle opening is 9.27%, airspeed is 240m/s, height is 10000m and attack angle is 0.26deg. The responses of airspeed, attack angle and height are shown in Figure 3. As can be seen from the results, the flight states almost keep the trim values, and have a little oscillating with long periods, caused by the small deviations of the given elevator angle and the throttle opening values.

When the elevator angle increment is 1deg, the flight states responses are shown in Figure 4. Although the aircraft system in trim state is stable, we get that the system without suitable control system is being to diffuse even if the elevator angle is slightly large.

When the throttle opening increment value is 0.01, the flight state responses are shown in Figure 5. We can see that when the airspeed slightly increases, the attack angle slightly decreases and the height increases when the throttle opening increases. If the throttle opening increases too much, this can also make the flight states diffuse.

3. DESIGNING OF CONTROL LAWS

3.1. Control System Structure

From the engineering perspective, in this work the classical control method is used to design the control laws. The control laws are designed step by step from the inner loop to the outer loop. A control structure diagram of the level change mode is established as shown in Figure 6. In Figure 6, the lateral control loop aims to achieve keeping the flight heading, and the throttle channel has a feedback of air speed to achieve keeping the air speed constant.

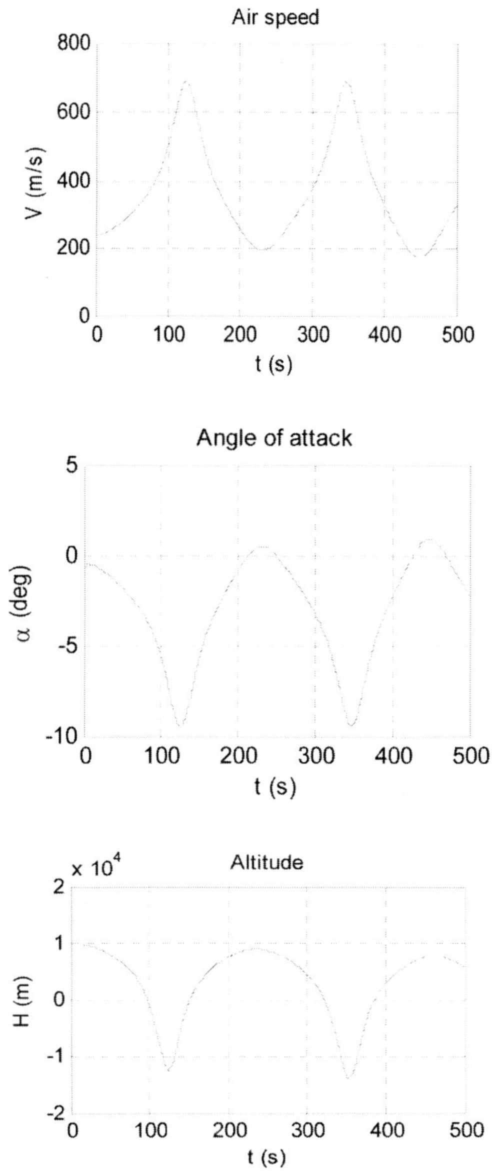

Figure 4. Flight state responses with elevator input

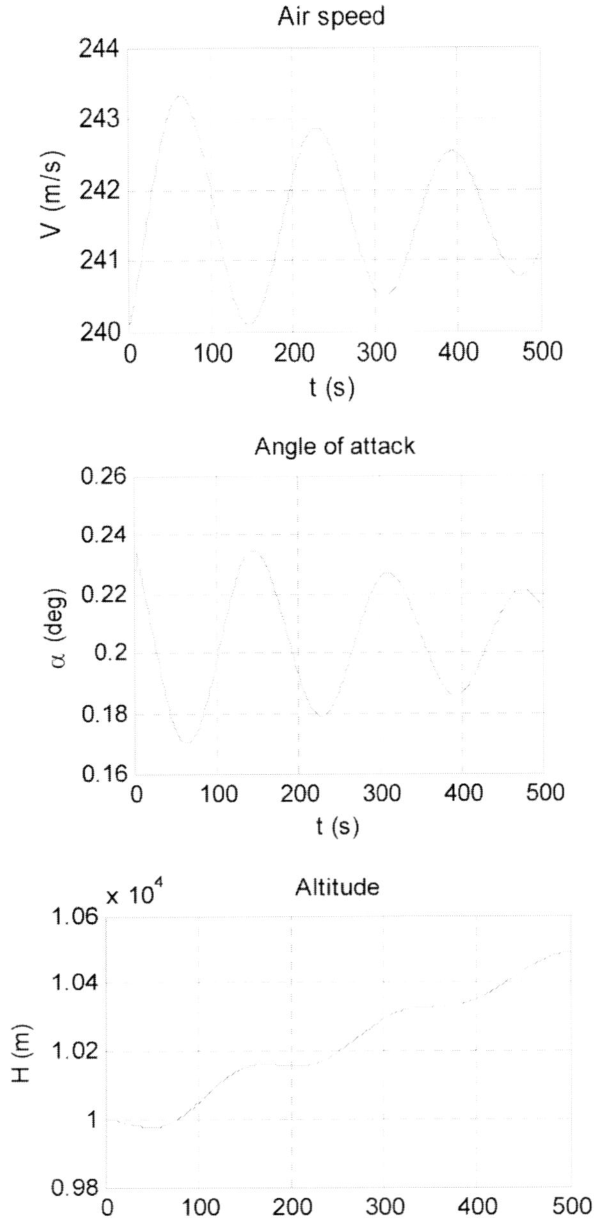

Figure 5. Flight state responses with throttle input

3.2. Design Of Control Laws

This work focuses on designing an elevator channel control law. The control structure of a longitude elevator channel is shown in Figure 7. In Figure 7, the inner loop is called damping augmentation loop, the outer loops are respectively called pitch angle control loop, vertical velocity control loop and altitude control loop. The design process of the throttle channel and the lateral channel are the same as for the elevator channel - here unnecessary details are no longer given.

The inner loop of the elevator channel uses the C∗C∗ control configuration, that is, through feedback of the normal overload nznz and the pitch angular rate q, it can achieve improvement of aircraft longitudinal flight qualities. The C∗C∗ signal is shown as (9), where VcoVco is cross velocity, according to engineering experience, its value is 120m/s.

$$C^* = n_z + \frac{V_{co}}{g} q \qquad (9)$$

Normal overload and angle of attack have a relationship expressed by

$$\Delta n_z = \frac{V_0}{g} Z_\alpha \Delta\alpha = 15.0122\Delta\alpha \qquad (10)$$

Then the C* signal can be expressed by

$$C^* = 15.0122\Delta\alpha + 12.4q \qquad (11)$$

The transfer function from the elevator deflection angle to the C* signal is shown as

$$G_{\Delta\delta_e}^{C^*}(s)=\frac{-23.8373s-42.1215}{s\left(s^2+0.8952s+2.6646\right)} \tag{12}$$

At first, the C∗C∗ controller must be designed and make C* index in the packet line [7], then use the root locus method to design the pitch angle control law - the whole pitch attitude control law of the elevator channel is shown as

$$\Delta\delta_e=\frac{-10}{s+10}[-0.42C^*+6.9(\Delta\theta_c-\Delta\theta)+3.3\int(\Delta\theta_c-\Delta\theta)dt] \tag{13}$$

The transfer function of the closed pitch attitude control loop is shown as

$$G_{theta\_cl}=G_{\Delta\theta_c}^{\Delta\theta}(s)$$
$$=\frac{21.48s^3+34.95s^2+18.86s+3.374}{0.1892s^6+2.171s^5+14.67s^4+54.26s^3+49.43s^2+18.86s+3.374} \tag{14}$$

In the case of keeping airspeed constant, that is with automatic throttle control, flight path bank angle increment Δμ and pitch angle increment Δθ have a relationship as shown in (15). Flight path bank angle increment and vertical velocity have a relationship as shown in (16).

$$\frac{\Delta\mu}{\Delta\theta}=\frac{-Z_\alpha}{s-Z_\alpha},\ Z_\alpha=-0.6130 \tag{15}$$

$$\frac{\Delta \dot{H}}{\Delta \mu} = \frac{V_0}{57.3} \tag{16}$$

From (15) and (16), the transfer function from ΔθcΔθc to ΔH˙ΔH˙ is expressed by

$$G_{hd\_open} = G_{\Delta\theta_c}^{\Delta \dot{H}}(s)$$
$$= \frac{3160s^3 + 5142s^2 + 2775s + 496.4}{0.1892s^7 + 2.287s^6 + 16s^5 + 63.25s^4 + 82.69s^3 + 49.16s^2 + 14.94s + 2.068} \tag{17}$$

Equation (17) is a seven-order transfer function which is not convenient for root locus design, so it can be converted to zero-pole form as

$$G_{hd\_open2} =$$
$$\frac{16702.63\,(s+0.5817)\,(s+0.5787)\,(s+0.4667)}{(s+5.404)(s+0.613)(s+0.58)(s^2+0.5208s+0.1466)(s^2+4.969s+38.81)} \tag{18}$$

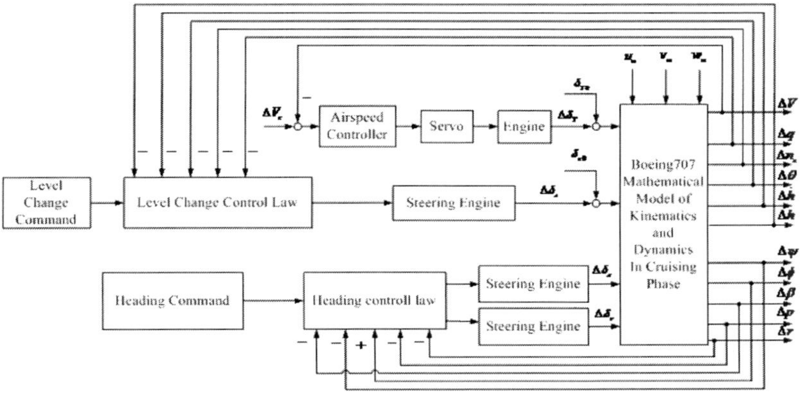

Figure 6. Control structure flowchart of the level change mode

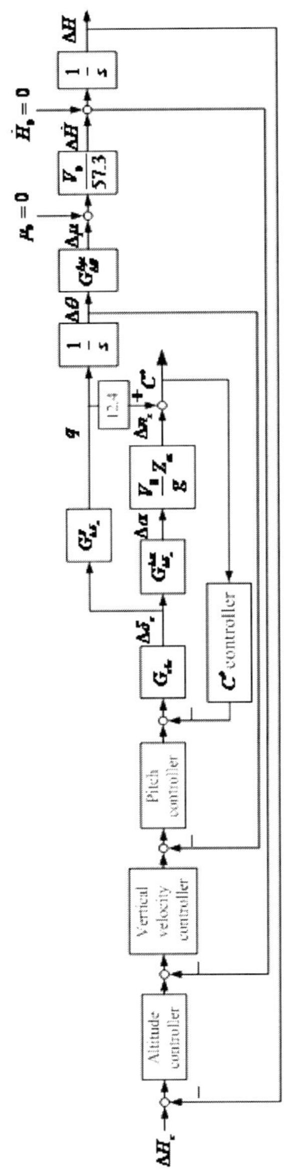

Figure 7. Control structure flowchart of the elevator channel

In (18), there are cancellations of zeros and poles, after eliminating two dipoles: P1=-0.58, Z1=-0.5787 and P2=-0.613, Z2=-0.5817, a simplified transfer function is obtained shown as

$$G_{hd\_open2} = \frac{16702.63\ (s+0.4667)}{(s+5.404)(s^2+0.5208s+0.1466)(s^2+4.969s+38.81)} \tag{19}$$

Based on (19), the vertical velocity controller can be designed to achieve tracking control of the vertical velocity command. Through the root locus method, parameters of the vertical velocity controller are designed as *P*=0.02, *I*=0.01. The altitude control is the outer loop of the vertical velocity control, when the parameter of the altitude controller is *P*=0.4, the root locus of the whole altitude control loop is shown as in Figure 8 - the poles of the closed loop are shown in Table 2.

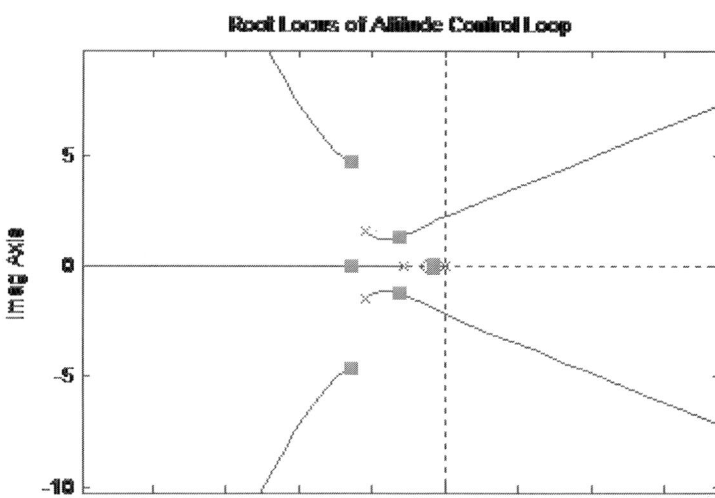

Figure 8. Root locus of the altitude control loop

Table 2 Poles of the altitude control closed loop

Pole Value	Damping	Frequency
$-0.297 \pm 0.0927i$	0.955	0.311
$-1.26 \pm 1.29i$	0.7	1.8
$-2.59 \pm 4.69i$	0.484	5.36

Finally, the designed control laws must be put into a nonlinear model to test the control effects. If the results are not good enough, the control parameters must be tuned again in a small range, until the best effects are achieved. This work fine-tunes the pitch attitude control law and all the control laws of the level change mode are shown as (20).

$$\begin{cases} \Delta\delta_e = \dfrac{-10}{s+10}(-5.2\Delta q - 0.42\Delta N_z + 8(\Delta\theta_c - \Delta\theta) + 2\int(\Delta\theta_c - \Delta\theta)dt) \\ \Delta\delta_T = \dfrac{20}{s+20}\dfrac{1}{s+1}(0.1(\Delta V_c - \Delta V) + 0.01\int(\Delta V_c - \Delta V)dt) \\ \Delta\delta_a = \dfrac{-10}{s+10}\left(-0.85\Delta p - 4\left(\Delta\phi - \dfrac{1}{s+1}\Delta\phi_c\right) - 0.8\int\left(\Delta\phi - \dfrac{1}{s+1}\Delta\phi_c\right)dt\right) \\ \Delta\delta_r = \dfrac{-10}{s+10}\left(-2\dfrac{s}{s+1}\Delta r + 6(\Delta\beta - \Delta\beta_c) + 3\int(\Delta\beta - \Delta\beta_c)dt\right) \end{cases} \quad (20)$$

Among this

$$\Delta\theta_c = -0.02(\dot{h} - \dot{h}_c) - 0.01\int(\dot{h} - \dot{h}_c)dt$$

$$\Delta V_c = \Delta\beta_c = 0$$

$$\Delta\dot{h}_c = -0.4(h - h_c)$$

4. SIMULATION AND ANALYSIS

Here we apply the designed control laws to the nonlinear model of the large aircraft to simulate and test. In simulation, from 35s to 45s, we input a gust disturbance, and decompose the speed of the gust into the X axis, Y axis and Z axis in the body coordinate. The speed component in each axis is respectively 5m/s, 3m/s and 2m/s. From 20s to 70s, with regard to altitude slope command the rate of climbing is 12m/s, this makes the flight level change from 9500m to 10100m. On other occasions, altitude holding command is connected into the altitude control loop. During the whole process, the airspeed holding command is connected to the throttle channel, and heading holding command is connected into the aileron channel. The simulation results are shown in Figure 9.

In Figure 9, altitude keeps constant at 9500m during the first 20 seconds, changes to 10100m from 20s to 80s, and finally keeps constant at 10100m. During the whole process, overshoot of altitude response is zero, gust disturbance has little effect on the level change mode, and the control system shows good anti-wind disturbance performance. In the climbing process, airspeed keeps constant, pitch angle increases to 3.2°3.2°, and its overshoot is less than 20%. Both in the connecting process and the cutting process of the level change mode, the angle of attack shows about two degrees of fluctuation, but in the steady state process, the angle of attack keeps constant. Lateral control achieves the action of heading holding, when the gust disappears, and the heading is the same as that before the gust appears. Roll angle changes within 10°10°, sideslip angle changes within 1°1°, and after the gust disappears, roll angle and sideslip angle soon restore to the initial trim state.

It is worth pointing out that when the flight level becomes much larger, because of the aerodynamic parameters, air density and temperature change, the control effect is obviously worse. Therefore, the control law needs to be designed again to adjust to the new flight environment, and smoothly switch to a control law for the new flight environment.

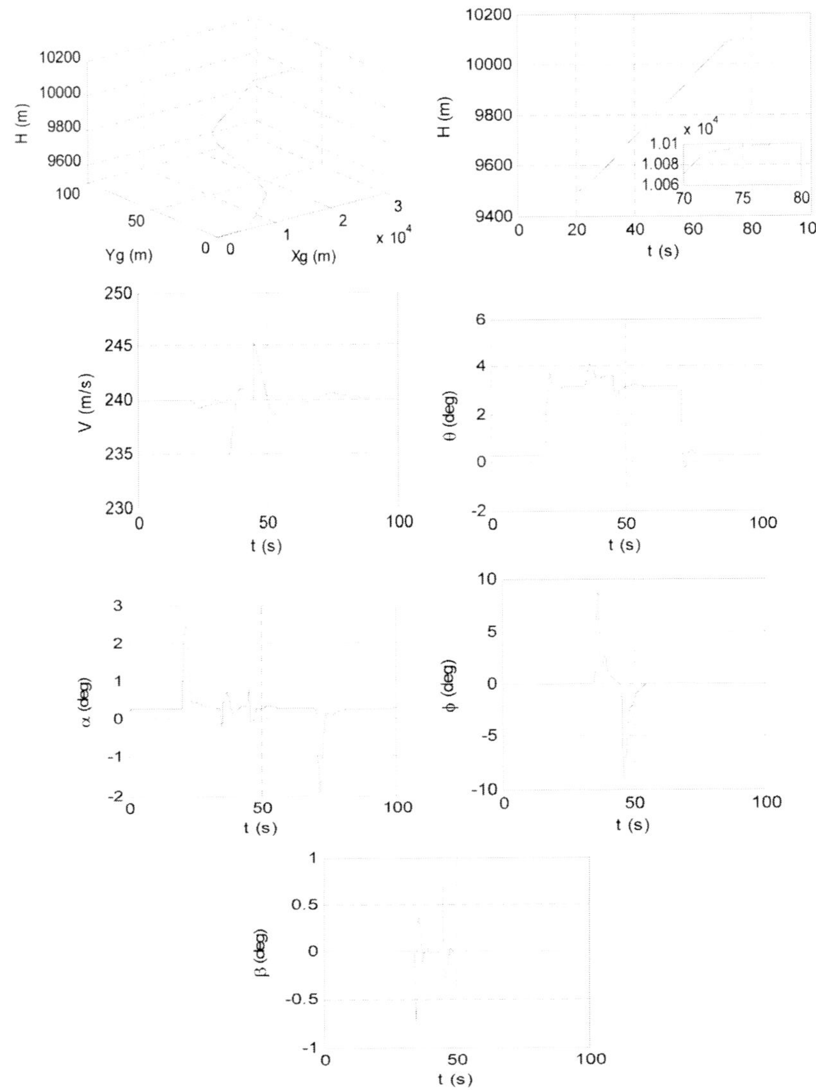

Figure 9. Nonlinear simulation results

5. CONCLUSION

This study builds up a nonlinear model of large aircraft based on the data of the Boeing 707 in cruising phase, Through the trim and linearization, a linear model of large aircraft is further obtained, and then the root locus method is used to design the control laws of Level Change mode of large aircraft. Based on the simplified transfer function, the longitudinal and lateral control laws are designed, to achieve airspeed holding, heading holding and level change. The nonlinear model based simulation results show that the designed control laws can meet the requirements of performance index of large aircraft, and the whole control system has a better anti-gust performance.

6. ACKNOWLEDGEMENTS

This study was supported by the NUAA Fundamental Research Funds (no. xlink013029, NN2012101, NP4003-56x1100), Aeronautical Science Foundation of China (no.2010ZA52002), Specialized Research Fund for the Doctoral Program of Higher Education (no. 20123218120015), and the Open Fund of Postgraduate Innovation Base of NUAA (no. kfjj20110207).

REFERENCES

1. Rini Akmeliawati, Iven Mareels. (2010) Nonlinear Energy-Based Control Method for Aircraft Automatic Landing Systems. IEEE Transactions on Control Systems Technology, vol. 18, no. 4, pp. 871-884.
2. Jong-Yeob Shin, Gregory Irene (2007) Robust Gain-Scheduled Fault Tolerant Control for a Transport Aircraft. Proceedings of

the IEEE International Conference on Control Applications, pp. 1209-1214.

3. Hakim Bouadi, Hondying Wu, Felix Mora-Camino (2011) Flight Path Tracking Based-on Direct Adaptive Sliding Mode Control. 2011 IEEE Intelligent Vehicles Symposium (IV), Baden-Baden, Germany.
4. Kato Akio, Inagaki Yoshiki (2008) Control Law for Automatic Landing Using Fuzzy-Logic Control. Transactions of the Japan Society for Aeronautical and Space Sciences, vol. 50, no. 170, pp. 274-283.
5. Jian Li, Weiguo Zhang, Guangwen Li (2005) Survey on the Application of μ Synthesis Method in Robust Flight Control System Design. Journal of Projectiles, Rockets, Missiles and Guidance. vol. 25, no. 4.
6. Leyao Li, Yan Li, Xinmin Wang. (2010) Design of Longitudinal Command Augmentation System for Large Civil Aircraft Relaxing Static Stability. Flight Dynamics, vol. 28, no. 4.
7. Rudolf Brockhaus. Flight Control. Springer, 1994.

CHAPTER 2

DIGITAL VIRTUAL FLIGHT TESTING AND EVALUATION METHOD FOR FLIGHT CHARACTERISTICS AIRWORTHINESS COMPLIANCE OF CIVIL AIRCRAFT BASED ON HQRM

Fan Liu, Lixin Wang, Xiangsheng Tan

School of Aeronautic Science and Engineering, Beihang University, Beijing 100191, China

ABSTRACT

In order to incorporate airworthiness requirements for flight characteristics into the entire development cycle of electronic flight control system (EFCS) equipped civil aircraft, digital virtual flight testing and evaluation method based on handling qualities rating method (HQRM) is proposed. First, according to HQRM, flight characteristics airworthiness requirements of civil aircraft in EFCS failure states are determined. On this basis, digital virtual flight testing model, comprising flight task digitized model, pilot controlling model, aircraft motion and atmospheric turbulence model, is used to simulate the realistic process of a pilot controlling an airplane to perform assigned flight tasks. According to the simulation results, flight characteristics airworthiness compliance of the airplane can be evaluated relying on the relevant regulations for handling qualities (HQ) rating. Finally, this method is applied to a type of passenger airplane in a typical EFCS failure state, and preliminary conclusions concerning airworthiness compliance are derived quickly. The research results of this manuscript can provide important theoretical reference for EFCS design and actual airworthiness compliance verification of civil aircraft.

KEYWORDS

Airworthiness verification; Civil aircraft; Flight characteristic; Flight control system; Handling qualities rating method (HQRM); Pilot model; Virtual flight testing

1. INTRODUCTION

Modern civil aircraft generally uses electronic flight control system (EFCS), which provides an electronic interface between pilot's controls and control surfaces, generating the actual surface commands for control about all three airplane axes. Since the existing airworthiness regulation for transport category aircraft

(CCAR-25-R4)[1] only targets at the airplane with mechanical flight control system,[2] for flight characteristics airworthiness compliance verification of civil aircraft in EFCS failure states, the specific condition about handling qualities rating method (HQRM) presented in Appendix 5 of Federal Aviation Administration (FAA) AC 25-7C must be adopted.2, 3 and 4 At present, flight characteristics airworthiness compliance verification of advanced civil aircraft such as A320, A330, A380, B777 and B787 in EFCS failure states has been completed based on HQRM.[5] In China, the flight characteristics airworthiness compliance verification of ARJ21-700 feeder liner in EFCS failure states will be undertaken for the first time through HQRM.

According to EFCS failure states, atmospheric disturbance levels and flight envelope conditions, HQRM defines the minimum handling qualities (HQ) level of civil aircraft which meets airworthiness requirements. Completion of assigned flight tasks is used to verify whether a civil aircraft meets the requirements of minimum HQ level so as to determine airworthiness compliance of the airplane. The verification method can be flight test or simulator test. Therefore, flight characteristics airworthiness compliance verification based on HQRM is a complex process of multi-factor coupling and pilot-aircraft closed system HQ rating.

By means of digital virtual flight testing technology, airworthiness compliance evaluation is integrated into the whole airplane development cycle, which represents a new concept for modern civil aircraft design.6, 7, 8, 9 and 10 Since flight characteristics of modern civil aircraft are closely related to the EFCS design, investigations on flight characteristics airworthiness compliance of civil aircraft at early design phases through digital virtual flight testing and evaluation method are of great importance. It can identify the EFCS design defects as soon as possible to avoid high cost associated with modifications at late design stages and to optimize the general aircraft development cycle. Furthermore, preliminary conclusions on flight characteristics airworthiness compliance can be obtained quickly and used to formulate more reasonable and efficient verification schemes for actual flight test and simulator test, to reduce flight risk and to avoid unnecessary verifications. Therefore, benefits such as improving the safety, reducing the cost and shortening the cycle of airworthiness verification can be obtained.

In this article, HQRM is first used to determine the airworthiness requirements for civil aircraft in EFCS failure states. On this basis, digital virtual flight testing and evaluation method for flight characteristics airworthiness compliance of civil aircraft is proposed based on HQRM. This method can be used for a quick preliminary evaluation of flight characteristics airworthiness compliance of civil aircraft in EFCS failure state, and to assist the EFCS design and the actual airworthiness compliance verification of civil aircraft in the whole development cycle.

2. FLIGHT CHARACTERISTICS AIRWORTHINESS REQUIREMENTS OF CIVIL AIRCRAFT IN EFCS FAILURE STATES

EFCS failure state, atmospheric disturbance level and flight envelope are the major factors that affect controllability and flight safety of civil aircraft. According to HQRM, flight characteristics airworthiness compliance of civil aircraft in EFCS failure states is determined by HQ levels of the airplane under the combinations of the above three factors.

A EFCS failure state, atmospheric disturbance level and flight envelope form a combination event, the occurrence probability of which is generally the product of occurrence probabilities of the three single events, shown as

$$X = X_c X_a X_e \quad (1)$$

where X is the occurrence probability of the combination event, X_c indicates the occurrence probability of the EFCS failure state, whose value is determined by the EFCS design, X_a is the occurrence probability of the atmospheric disturbance level, whose values of 10^0, 10^{-3} and 10^{-5} represent light, moderate and severe disturbances, respectively, X_e is the occurrence probability of the flight envelope, whose values of 10^0, 10^{-3} and 10^{-5} represent normal, operational and limit flight envelope, respectively. The combination events with the occurrence probability of X less than 10^{-9} (once per 10^9 flight hours) are not required to be carried out airworthiness verification, while those having an occurrence probability of X higher than 10^{-9} and conforming to the practical flight conditions must be verified.

Flight characteristics airworthiness compliance of civil aircraft in EFCS failure states are expressed as follows: for all combination events to be verified, the airplane should at least meet the requirements of minimum HQ level shown in Table 1, where SAT, ADQ and CON stand for satisfactory, adequate and controllable, respectively.

Table 1. Requirements of minimum handling qualities level.

Flight condition ($X_c X_e$)	Atmospheric disturbance level (X_a)								
	Light level			Moderate level			Severe level		
	Normal	Operational	Limit	Normal	Operational	Limit	Normal	Operational	Limit
Probable	SAT	SAT	ADQ	ADQ	CON	CON	CON	CON	CON
Improbable	ADQ	ADQ	CON	CON	CON		CON		

For any combination event, according to the combined probability $X_c X_e$ of the EFCS failure state and the flight envelope, the flight condition under this combination event is "Probable Condition" (X_cX_e in the range of 10^0–10^{-5}) or "Improbable Condition" (X_cX_e in the range of 10^{-5}–10^{-9}). At the same time, in accordance with the atmospheric disturbance level of this combination event, the airworthiness requirements of minimum HQ level can be determined on the basis of Table 1.

As shown in Table 1, HQ level is classified into "Satisfactory", "Adequate" and "Controllable". FAA definition and the corresponding Military Standard lever and quality of each HQ level are shown in Table 2.

Table 2. FAA definition and the corresponding Military Standard lever and quality of each HQ level.

HQ level	FAA definition	Military Standard	
		Level	Quality
Satisfactory	Full performance criteria met with routine pilot effort and attention	1	Satisfactory
Adequate	Adequate for continued safe flight and landing, full or specified reduced performance met, but with heightened pilot effort and attention	2	Acceptable
Controllable	Inadequate for continued safe flight and landing, but controllable for returning to safe flight condition, a safe flight envelope, and /or reconfiguration so that HQ is at least adequate	3	Controllable

For HQ rating by digital virtual flight testing and evaluation method, the FAA definitions of HQ level are on a qualitative basis, while some criteria of Military Standard would be the quantitative basis if necessary. In addition, HQ rating may also refer to some clauses of Airworthiness Regulation CCAR-25-R4.

Therefore, HQ rating basis of civil aircraft in EFCS failure states mainly includes the following four aspects:

1 The ability to complete the assigned flight tasks. The airplane should be able to complete the assigned flight tasks and meet the relevant performance criteria of Airworthiness Regulation, Military Specification or other standards.

2 The control and state variables of the airplane. The aircraft control and state variables should comply with various constraints during the performing of flight tasks, such as velocity limit, angle of attack limit and control surface deflection limit.

3 The control force and control force gradient of the pilot. The pilot control force and control force gradient should meet the requirements of Airworthiness Regulation and Military Specification, such as the maximum control force of wheel in the Clause 25.143(d) of CCAR-25-4R, and the maximum control force gradient for longitudinal maneuvering task in the Clause 3.2.2.2 of MIL-F-8785C.[11]

4 The pilot controlling model. In order to ensure that digital virtual flight testing fully represents the actual flight test or simulator test, the pilot controlling model should reflect the actual behavioral characteristics of pilot control.

3. DIGITAL VIRTUAL FLIGHT TESTING AND EVALUATION METHOD FOR AIRWORTHINESS COMPLIANCE

For flight characteristics airworthiness compliance evaluation of civil aircraft in EFCS failure states, first airworthiness requirements should be determined according to HQRM, including all combination events to be verified and the corresponding requirements of minimum HQ level, as described above in Chapter 2. On this basis, for each combination event to be verified, digital virtual flight testing is used to simulate the realistic process of a pilot controlling the airplane to perform the assigned flight tasks, and the airplane actual HQ level is evaluated in terms of the HQ rating basis. Thus, airworthiness compliance conclusion is obtained by comparing the actual HQ level with the minimum HQ level which meets airworthiness requirement.

Therefore, targeting at flight characteristics airworthiness compliance of civil aircraft in EFCS failure states, digital virtual flight testing and evaluation method mainly includes flight task digitized model, pilot controlling model, aircraft motion and atmospheric disturbance model, and digital virtual flight testing and airworthiness compliance evaluation, as shown in Fig. 1.

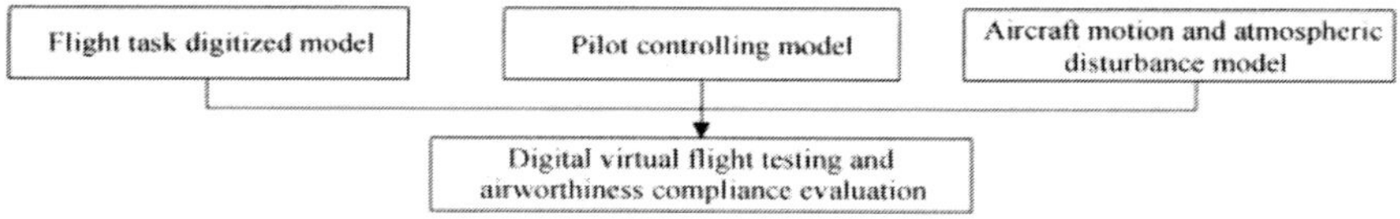

Figure. 1. Framework of digital virtual fight testing and evaluation method.

3.1. Fight task digitized model

Flight tasks for HQ rating of civil aircraft fall into three categories, as shown in Table 3.[3]"Trim & unattended operation" is mainly to verify the characteristics of an airplane to stay at or depart from an initial trim or unaccelerated condition, which will be appropriate for the failures affecting the trim characteristics of the airplane. "Closed-loop precision tracking of flight path" is generally the pilot closed-loop task performed in routine commercial flight. This type of task will be suitable for the failures affecting the pilot closed-loop tracking performance of the airplane. "Large amplitude maneuvering" basically represents stability and control tests of flight characteristics airworthiness verification, which will apply to the failures affecting stability and control of the airplane.

Table 3. Flight task for HQ rating.[3]

Category	Typical fight task
Trim & unattended operation	Dynamic and flight-path response to pulse (3 axes)
Closed-loop precision tracking of flight path	Crosswind landing
Large amplitude maneuvering	Pitch: symmetric pull-up/push-over
	Roll: rapid bank-to-bank roll
	Yaw: sudden heading change
	Operational: climbing/diving turn

The flight tasks available for choice are not limited to Table 3. In principle, the selected tasks should fully reflect the impact of EFCS failures on the control and flight safety of the airplane under evaluation.

Flight task digitized model is to generate digital flight commands based on the requirements of the flight task.[12] Take "Crosswind Landing" as an example, the requirements of this task are written as:[13]

1 At approach stage, the airplane should glide along the runway centerline at a certain glide angle, and keep a certain velocity.

2 When the flight altitude is lower than a certain value, the airplane should be flared to enable a mild touchdown at a smaller descent rate.

3 Before the touchdown, the roll and yaw attitudes of the airplane should be controlled within the safe range.

On this basis, the flight commands for "Crosswind Landing" of a certain passenger airplane are written as

$$\begin{cases} V_c = 1.23V_{SR} \\ \gamma_c = \begin{cases} -3^\circ & h > 6 \\ 0^\circ & h \leqslant 6 \end{cases} \\ y_c = 0 \\ \psi_c = \begin{cases} -\arctan(V_w/V) & h > 3 \\ 0^\circ\text{–}5^\circ & h \leqslant 3 \end{cases} \end{cases} \tag{2}$$

where V_c, γ_c, y_c and ψ_c represent the commands of velocity, glide angle, lateral position and yaw angle, respectively; V_{SR} is the stall speed, V_w the crosswind speed, $V$the velocity. Eq. (2) does not contain the roll angle command. This is because under the constraint conditions of Eq. (2), the roll angle is determined by the yaw angle.

3.2. Pilot controlling model

The pilot controlling model is to generate the inputs of throttle and control surfaces according to the flight commands and the feedback of airplane state outputs. The pilot's control behavior of civil aircraft is generally divided into throttle control, longitudinal control, lateral control and directional control, which control the velocity, pitch motion, roll motion and yaw motion of airplane, respectively. For the velocity control, longitudinal control, lateral control and directional control, single-channel pilot controlling model is shown as Fig. 2.[14]

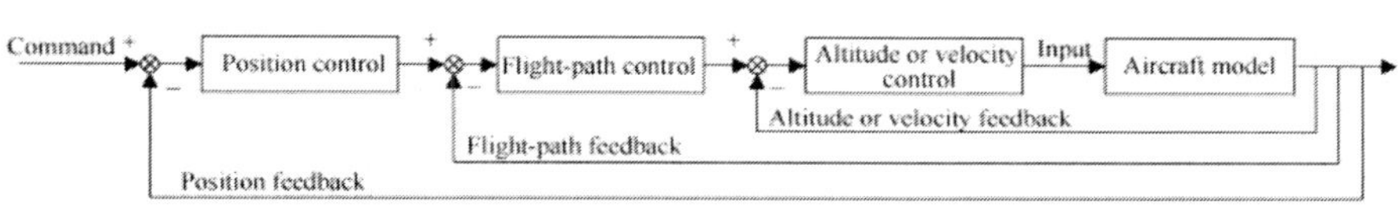

Figure. 2. Single-channel pilot controlling model.[14]

As shown in Fig. 2, single-channel pilot controlling mode consists of outer loop for position control, intermediate loop for flight-path control and inner loop for attitude or velocity control. The structure in Fig. 2 is the pilot controlling model with position as flight command. With attitude or velocity control loop retained only, the pilot controlling model with attitude or velocity as flight command is obtained, while with both flight-path control loop and attitude control loop, the pilot controlling model with flight-path as flight command.

For single-channel pilot controlling model, both the outer loop and intermediate loop simply adopt proportional control in order to convert the error signal to the control signal of the inner loop. For the sake of eliminating the steady-state error, the outmost loop can also adopt proportional–integral controller (PI controller). The outmost loop means the intermediate loop when the outer loop does not exist. The inner loop adopts Hess structural pilot model,[15] as shown in Fig. 3, to yield the crossover model of the human pilot.[16] K_m and K_r represent the gains on the output error and the difference $\dot{M}_c - \dot{M}_c$ respectively, and G_{nm} the pilot's limb neuromotor dynamics. Refs. 15 and 17 have given the gain scheduling methods for K_m and K_r and presented the expression of G_{nm}.

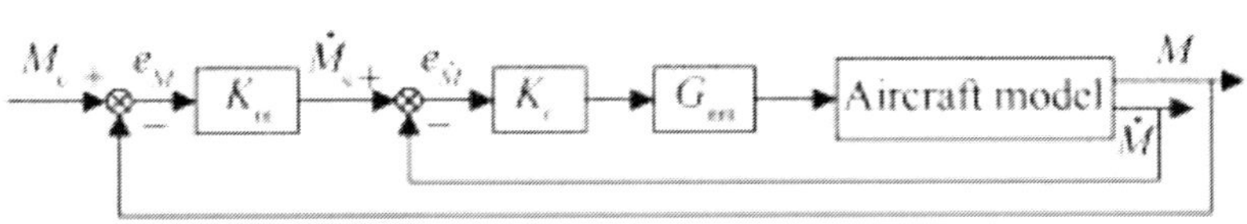

Figure. 3. Hess structural pilot model.[15]

The crab method of "Crosswind Landing" is used as an example in the following analysis. According to the flight commands for "Crosswind Landing" shown in Eq. (1), a multi-channel pilot controlling model consisting of throttle channel, pitch channel, roll channel and yaw channel is required to be established to control velocity, slide angle, lateral position and yaw attitude of the airplane, respectively, as shown in Fig. 4, where θ, ϕ, ψ and χ represent the pitch attitude, roll attitude, yaw attitude and head angle, respectively; Y_{PV}, $Y_{P\theta}$, $Y_{P\phi}$ and $Y_{P\psi}$ represent the transfer functions of inner loops of pilot control model in each channel, respectively; $Y_{P\gamma}$ and $Y_{P\chi}$ are the transfer functions of intermediate loop of pilot control model in pitch channel and roll channel, respectively; Y_{Py} is the transfer function of outer loop of pilot control model in roll channel.

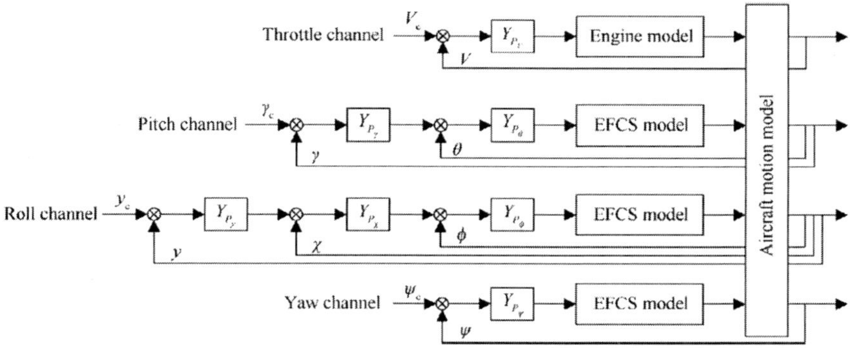

Figure. 4. Multi-channel pilot controlling model for "Crosswind Landing".

The throttle channel uses velocity as flight command. The pitch channel uses flight-path angle as flight command, which is responsible for flight-path control loop and pitch attitude control loop. The roll channel uses lateral position as flight command, which is responsible for lateral position control loop, heading control loop and roll attitude control loop. The yaw channel directly uses yaw attitude as flight command.

3.3. Aircraft motion and atmospheric disturbance model

3.3.1. Aircraft motion model

Aircraft motion model includes airframe aerodynamic model and EFCS model. In order to build the digital virtual flight testing model based on simulation of "Task–Pilot–Aircraft–Environment", airframe aerodynamic model should be able to reflect the influence of atmospheric disturbance on the motion of the airplane. On the foundation of six-freedom aircraft motion equation in quiet atmosphere,[18] velocity, angle of attack, sideslip angle and aerodynamic moment coefficients are corrected, as shown by Eqs. (3) and (4). Thus, the airframe aerodynamic model in turbulent atmosphere can be established.

$$\begin{cases} \Delta V = -u_{\mathrm{w}} \\ \Delta a = -w_{\mathrm{w}}/V \\ \Delta \beta = -v_{\mathrm{w}}/V \end{cases} \tag{3}$$

$$\begin{cases} \Delta C_l = [C_{lp}(p - \omega_{\mathrm{w}y}) + C_{lr}(r - v_{\mathrm{w}x})]b/2V \\ \Delta C_m = [C_{mq}(q - \omega_{\mathrm{w}x}) + C_{m\dot{a}}(\dot{\alpha}_k - \omega_{\mathrm{w}x})]c/2V \\ \Delta C_n = [C_{np}(p - \omega_{\mathrm{w}y}) + C_{nr}(r - v_{\mathrm{w}x})]c/2V \end{cases} \tag{4}$$

Ref. [19] has given the correction principle and the symbol meanings in Eqs. (3) and (4).

For the EFCS equipped aircraft, the pilot control forces are converted into electric signals, which are sent to the flight control computer

(FCC) through wires. Then the FCC performs a calculation based on the pre-configured flight control law and sends electric signals to the control surface actuators. Finally the control surfaces begin to move and the aircraft will achieve the desired flight attitude, flight-path or position. Noteworthy, the FCC can calculate and send electric signals to the control surface actuators directly under the mode of automatic flight control. According to the operation principle of EFCS, the basic EFCS model is shown as Fig. 5.

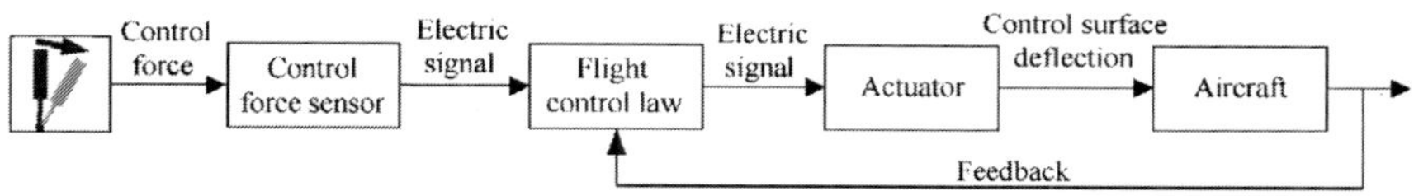

Figure. 5. Basic EFCS model.

The control force sensor is the control column or control pedal force sensor model which converts the pilot control force into electric signal. Flight control law is the pre-configured algorithm in FCC. The modeling of control force sensor, flight control law and actuator should be based on the actual EFCS design of civil aircraft.

The failure of EFCS is described by failure mode, failure impact and failure probability.[4] EFCS failures of civil aircraft generally involve pitch, roll and yaw control malfunction, and the specific failure modes include jam, floating, incorrect action, gain variation, rate variation, trim function yaw failure, and so on. The definitions and impacts of typical failure modes are shown in Table 4.

The main impacts of EFCS failures are to reduce the control effectiveness and trim capacity of the airplane. In addition, some failures may cause adverse force and moment that impact the airplane control.

Table 4. Typical failure modes.

Failure mode	Definition	Main impact
Jam	Control column, control pedal or control surface gets stuck in the neutral or normal position	
Floating	Control surface floats to the position where aerodynamic hinge moment keeps zero	Weakening control effectiveness; generating undesirable force and moment
Incorrect action	Control surface acts without instruction	
Gain abnormity	Gain of control surface changes abnormally	Deteriorating handling qualities
Rate abnormity	Control surface acts faster than normal rate	Increasing the sensitivity of pilot control
Trim function failure	Trim function of control surface failures in the neutral or normal position	Heightening pilot effort and attention

According to the flight control system design requirements, the EFCS failures' occurrence probabilities are associated with their impacts, namely, the occurrence probability of failure with severe impact ought to be low. HQRM determines the airworthiness requirements of the minimum HQ level according to the occurrence probability of combination events, as shown in Table 1. Therefore, as a factor of combination event occurrence probability, the EFCS failure occurrence probability will impact on the airworthiness requirements of minimum HQ level, namely, the higher the occurrence probability of EFCS failure, the higher the airworthiness requirements of minimum HQ level.

3.3.2. Atmospheric disturbance model

For the flight characteristics airworthiness verification of civil aircraft according to HQRM, atmospheric disturbance required to be considered mainly includes gust, wind shear and crosswind, as shown in Table 5.[3]

The model of gust and wind shear[11] are shown in Eqs. (5) and (6), respectively.

$$V_w = \begin{cases} 0 & x < 0 \\ \frac{V_m}{2}\left(1 - \cos\frac{\pi x}{d_m}\right) & 0 \leqslant x \leqslant d_m \\ V_m & x > d_m \end{cases} \quad (5)$$

Table 5. Atmospheric turbulence for airworthiness verification.[3]

Atmospheric disturbance	Disturbance intensity (m/s)		
	Light level	Moderate level	Severe level
Gust	≤0.91	0.91–2.44	2.44–6.10
Wind shear	≤5.14	5.14–12.86	12.86–23.15
Cross wind	≤5.14	5.14–12.86	12.86–23.15

$$u_w = u_{20} \frac{\ln(h/Z_0)}{\ln(20/Z_0)} \tag{6}$$

Ref. [11] has described the model of gust and wind shear in detail and given the meanings of the symbols in Eqs. (5) and (6). Since crosswind is the component of a constant wind that is blowing across the runway, the wind speed can stand for its model.

3.4. Digital virtual flight testing and airworthiness compliance evaluation

According to the flight task digitized model, pilot controlling model, aircraft motion model (including airframe aerodynamic model and EFCS model) and atmospheric disturbance model described above, the digital virtual flight testing model is established as shown inFig. 6.

For each combination event to be verified, firstly, suitable flight tasks are selected then digitized by the flight task digitized model. After that, the initial flight conditions are set, including flight height, velocity, configuration, center of gravity, state of EFCS of the airplane as well as the wind field, according to the combination event. Finally, the realistic process of pilot performing the flight tasks is facsimiled via the digital virtual flight testing and the relevant data are collected for HQ level evaluation in terms of the HQ rating basis described in Section 2.

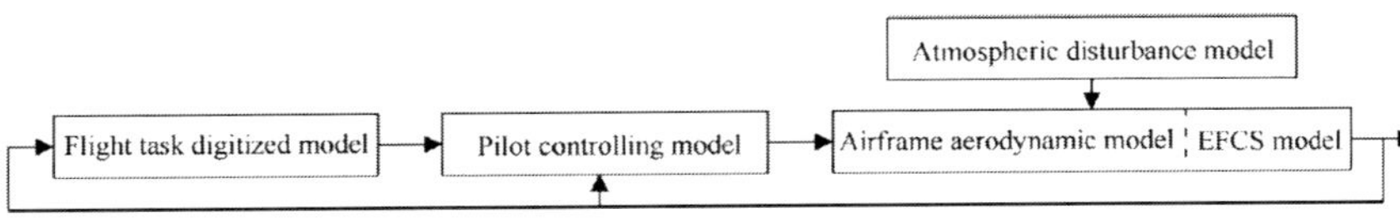

Figure. 6. Digital virtual flight testing model.

For specific combination event, when the evaluating result of HQ level is equal to or better than the corresponding minimum HQ level requirement in Table 1, it is thought that flight characteristics of the airplane under this combination event meet airworthiness requirements.

4. EXAMPLE

An example of digital virtual flight testing and evaluation for flight characteristics airworthiness compliance of a certain EFCS equipped passenger airplane will be presented. The EFCS failure state is "Control authority degrading of rudder" with the occurrence probability of 5.1×10^{-6}.

"Control authority degrading of rudder" refers to the value of rudder gain at low velocity becoming the value at high velocity. For the example airplane, the rudder gain will decrease with the increasing of velocity. As a result, the deflection range of rudder will reduce due to this failure. (−20° to 20°) is the maximum deflection range and (−10° to 10°) is the deflection range at the highest velocity. In order to verify the most serious failure condition, (−10° to 10°) will be the rudder control authority of the example airplane in this EFCS failure state.

Through calculation and analysis, the airworthiness requirements (including all combination events to be verified and the corresponding requirements of minimum HQ level) and the flight tasks for the example airplane are shown in Table 6.

Table 6. Airworthiness requirements and flight tasks for the example airplane.

Case	Atmospheric disturbance	Flight envelope	Combination event probability	Flight task	Requirement of minimum HQ level
A	Light	Normal	5.1×10^{-6}	Crosswind Landing	Adequate
B	Moderate	Normal	5.1×10^{-9}	Crosswind Landing	Controllable
C	Light	Operational	5.1×10^{-9}	Crosswind Landing	Adequate

"Crosswind Landing" is a pilot closed-loop task required to be performed in routine commercial flight and should be examined in normal flight envelope.[3] Therefore, only Cases A and B in Table 6 need to be verified. For "Crosswind Landing", the indexes of HQ rating are shown in Table 7.

Table 7. Indexes of HQ rating for "Crosswind Landing".

Case	Requirement of minimum HQ level	HQ rating index	Basis
A	Adequate	1. Finishing the task of crosswind landing 2. Both the roll and yaw attitude are less than 6° when touch down 3. The wheel control force and pedal control force is less than 222 N and 667 N, respectively	Appendix 5 of Ref. [3,20] Clause 25.143 of Ref. [1]
B	Controllable	Controllable for returning to safe flight condition, a safe flight envelope, and/or reconfiguration so that HQ is at least adequate	Appendix 5 of Ref. [3]

According to the characteristics of the example airplane, the flight commands for "Crosswind Landing" are shown as Eq. (2); the pilot controlling model is shown as Fig. 4; the initial flight conditions of digital virtual flight testing are shown in Table 8.

Table 8. Initial flight conditions for "Crosswind Landing".

Case	Configuration	Crosswind speed (m/s)	Velocity	Altitude (m)	Weight condition (adverse)
A	Landing	−5.14	$1.23V_{SR}$	100	Maximum landing weight, maximum inertia moment, forward center of gravity
B	Approaching	−12.86	$1.23V_{SR}$	100	

By means of digital virtual flight testing, the time histories of control and state variables of the example airplane performing "Crosswind Landing" are shown in Fig. 7 and Fig. 8, respectively. Since the failure "control authority degrading of rudder" primarily influences the lateral–directional control of the example airplane, the simulation results mainly include the lateral-directional control and state variables are given.

As shown in Fig. 7, F_a and F_r represent the control forces of wheel and pedal, respectively, δ_a and δ_r are the deflections of aileron and rudder, respectively. As shown in Fig. 8, H, Y and V represent the flight altitude, lateral position and velocity.

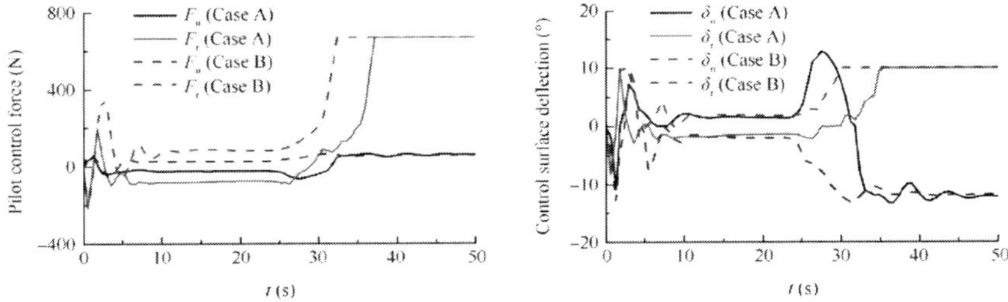

(a) Pilot control force

(b) Control surface deflection

Figure. 7. Control variables of "Crosswind Landing".

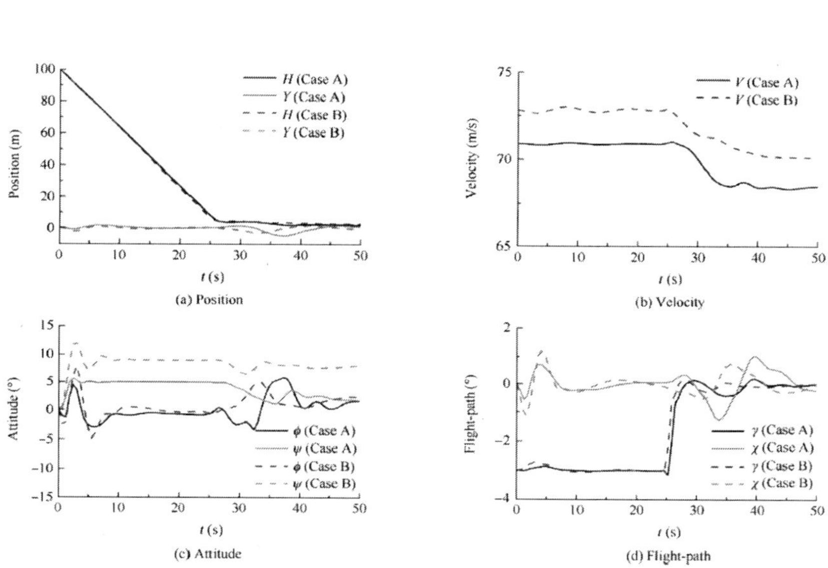

Figure. 8. State variables of "Crosswind Landing".

From the simulation curve, it can be seen that the whole flight process is divided into three stages: initial disturbed motion, steady approach and flaring before touchdown.

Since the failure does not produce any force and moment, the example airplane is only subjected to the crosswind disturbance at the initial stage: the airplane rolls to left under the action of the right crosswind and the heading deviates from the runway centerline (where $Y = 0$) and drifts left (the initial 2 s in Fig. 8(a), (c) and (d)). In order to resist the disturbance by crosswind, the pilot has to apply upwind ailerons to control the airplane to roll to the right, thus to eliminate the heading deviation; the upwind rudder is simultaneously applied to control the aircraft to yaw to right, therefore to turn the nose pointing to the direction of incoming flow (the 2–5 s in Fig. 7, the airplane maintains a steady approach during the next 20 s, and keeps the heading along with the runway centerline (see Fig. 8(d)). When the flight altitude is lower than 6 m, the pilot flatters out the airplane to reduce the descent rate (see Fig. 8(a)). When the flight height is lower than 3 m, downwind rudder is applied to reducing the yaw angle, with a simultaneous application of upwind ailerons to counteract the airplane left rolling trend caused by the right crosswind (the last 25 s in Fig. 7). Finally, the example airplane is grounded near the runway centerline (see Fig. 8(a)).

According to the terms of the HQ rating basis described in Chapter 2. First, as shown inFig. 8, in both Cases A and B, the example airplane can finish the flight task of "Crosswind Landing". Second, the control and state variables of the airplane comply with the constraints, except for the yaw angle exceeding the safety range upon touchdown in Case B, as shown in Fig. 8(c) and Table 7. Third, for the Cases A and B, during the whole flight process, the wheel control forces are less than 167 N, and the pedal forces are less than 667 N (see Fig. 7(a)), which meet the requirements on control force shown in Table 7. Finally, the selection of pilot controlling model gains are carried out according to Ref.[17], which can reflect the actual behavioral characteristics of the pilot. Pilot controlling model gains are shown in Table 9, where K_{p1} and K_{i1} represent the proportional gain and integral gain of PI controller in outer loop, respectively; K_{p2} and K_{i2} the proportional gain and integral gain of PI controller in intermediate loop, respectively. As shown in Table 9, the element G_{nm} is a second-order system model.

Table 9. Pilot controlling model gains.

Channel	Outer loop		Intermediate loop		Inner loop		
	K_{p1}	K_{i1}	K_{p2}	K_{i2}	K_m	K_r	G_{nm}
Throttle					0.62	0.20	$\frac{100}{s^2 + 2 \times 0.707 \times 10 \times s + 100}$
Pitch			1.5	1	2.12	2.36	
Roll	0.004	0.00125	5.3		1.98	0.52	
Yaw					1.72	−5.22	

Therefore, the airplane actual HQ in Case A acquires level of "sufficient", which meets the airworthiness requirement. The airplane actual HQ in Case B does not acquire level of "controllable" and fails to meet the airworthiness requirement.

According to the digital virtual flight testing and evaluation results above, suggestions about the design and the actual airworthiness compliance verification of the example airplane are proposed and stated as follows.

1 Increase the rudder size of the example airplane appropriately to make up for the diminution of yaw control effectiveness due to the failure.

2 Reduce the occurrence probability of "Control authority degrading of rudder" by improving the system reliability of the EFCS to avoid the occurrence of Case B.

3 The digital virtual flight testing and evaluation results can provide references for the verification scheme formulation and the pilot's controlling strategy identification of the actual flight test and simulator test in the verification process.

5. CONCLUSIONS

1 For flight characteristics airworthiness compliance of civil aircraft in EFCS failure states, digital virtual flight testing and evaluation method is proposed based on HQRM. First, we determine the airworthiness requirements for flight characteristics and HQ rating basis for airworthiness compliance evaluation according to HQRM. On this basis, we build a digital virtual flight testing model for simulating the realistic process of a pilot controlling an airplane to perform assigned flight tasks.

Finally, we rate the HQ level and evaluate the airworthiness compliance according to simulation results and HQ rating basis.

2 The research results of this manuscript can provide valuable theoretical references for EFCS design and actual airworthiness compliance verification of civil aircraft: airworthiness requirements can be incorporated into the entire development cycle of civil aircraft, and preliminary conclusions on airworthiness compliance can be obtained quickly, which is valuable for verifying and improving the EFCS design, formulating efficient scheme of the actual flight test and simulator test, as well as identifying riskless pilot controlling strategy in the verification process.

ACKNOWLEDGEMENTS

The authors are grateful to Dr. Xu Dongsong for joint research. They also thank the anonymous reviewers for their critical and constructive review of the manuscript. This study was supported by the National High-tech Research and Development Program of China (No. 2014AA110500).

REFERENCES

1 Civil Aviation Administration of China. China civil aviation regulations part 25 airworthiness standards of transport category aircraft. Beijing: Civil Aviation Administration of China; 2011. p. 1–28 [Chinese].

2 Zhu L, Huang MY, Ou XP, Song ZT. Analysis of key technologies of airworthiness certification of electronic flight control systems for civil transport airplane. CGNCC 2010: 2010 Chinese guidance,

navigation, and control conference. Shanghai: Science Press; 2010 [Chinese].

3 Federal Aviation Administration. Flight test guide for certification of transport category airplanes. Washington, DC: Federal Aviation Administration; 2012. p. A5-1–10.

4 Zhang YN, Li Y, Jin L. Airworthiness verification of handling qualities of the aircraft with electronic flight control system. Flight Dyn 2012;30(2):117–20 [Chinese].

5 Zhang YN, Li Y, Gao YK. Commercial transport aircraft flight simulation flying qualities airworthiness verification. Appl Mech Mater 2012;235:170–5.

6 Mendonca CB, Silva ET, Curvo M, Trabasso LG. Model-based flight testing. J Airc 2013;50(1):176–86.

7 7. Xu HJ, Liu DL, Xue Y, Zhou L, Min GL. Airworthiness compliance verification method based on simulation of complex system. Chin J Aeronaut 2012;25(5):681–90.

8 Schari J, Mavris DN, Burdun IY. Use of flight simulation in early design: formulation and application of the virtual testing and evaluation methodology. 2000. Report No.: AIAA-2000-5590.

9 Burdun IY, DeLaurentis DA, Mavris DN. Modeling and simulation of airworthiness requirements for an HSCT prototype in early design. 1998. Report No.: AIAA-1998-4936.

10 Hines DO. Simulation in support of flight testing. 2000. Report No.: RTO-AG-300.

11 U.S. Department of Defense. Flight qualities of piloted aircraft. Arlington: U.S. Department of Defense; 1997. p. 679–84.

12 Robert KH. Use of a task-pilot-vehicle (TPV) model as a tool for flight simulator math model development; 2010. Report No.: AIAA-2010-7620.

13 Federal Aviation Administration. Airplane flying handbook. Washington, DC: Federal Aviation Administration; 2004. p. 813–5.

14 Hosman R, Schuring J, Geest P. Pilot model development for the manual balked landing maneuver. 2005. Report No.: AIAA-2005-5884.

15 Hess RA. Simplified approach for modeling pilot pursuit control behavior in multi-loop flight control tasks. J Aerosp Eng 2006; 220(G2):85–102.

16 McRuer DT, Jex HR. A review of quasi-linear pilot models. IEEE Trans Hum Factors Electron 1967;8(3):231–49.

17 Hess RA. Obtaining multi-loop pursuit-control pilot models from computer simulation. 2007. Report No.: AIAA-2007-247.

18 Fang ZP, Chen WC, Zhang SG. Aircraft flight dynamics. 1st ed. Beijing: Beihang University Press; 2005. p. 174–81 [Chinese].

19 Xiao YL, Jin CJ. The principle of flight in atmospheric disturbance. 1st ed. Beijing: National Defense Industry Press; 2002. p. 82–4 [Chinese].

20 Airbus. Flight operations briefing note landing techniques [Internet]. 2008 Mar [cited 2014 Feb 21]. Available from: http://www.airbus.com/fileadmin/media_gallery/files/safety_library_items/AirbusSafetyLib_-FLT_OPS-LAND-SEQ05.pdf.

CHAPTER 3

GENETIC PROGRAMMING IN APPLICATION TO FLIGHT CONTROL SYSTEM DESIGN OPTIMISATION

Anna Bourmistrova[1] and Sergey Khantsis[1]

[1] Royal Melbourne Institute of Technology, Australia

1. INTRODUCTION

1.1. Evolutionary Algorithms

EAs are often considered as an example of *artificial intelligence* and a *soft computing* approach. Their unique ability to search for complete and global solutions to a given problem makes EAs a powerful problem solving tool which combine such important characteristics as robustness, versatility and simplicity.

Historically, there exist several branches of EAs, namely Genetic Algorithms, Genetic Programming, Evolutionary Programming and Evolutionary Strategies. Their development started independently in the 1960s and 70s. Nevertheless, all of them are based on the same fundamental principle - evolution. 'Evolution' is used here in its Darwinian sense, the advance through 'survival of the fittest'.

Despite of a remarkable simplicity, EAs have proven to be capable of solving many practical tasks. The first and obvious application is numerical optimisation (minimisation or maximisation) of a given function. However, EAs are capable of much more than function optimisation or estimation of a series of unknown parameters within a given model of a physical system. Due to, in a large part, their stochastic nature, EAs can *create* such complex structures as computer programs, architectural designs and neural networks. Several applications of EAs have been known to produce a patentable invention (Koza et al., 1996, Koza et al., 1999 and Koza et al., 2003).

Unfortunately, such a truly intelligent application of EAs is rarely used for practical purposes. GP and similar algorithms often require a supercomputing power to produce an optimal solution for a practical task. This may be overcome, at least partially, by narrowing the search space.

A general engineering design practice is to propose a new design based on existing knowledge of various techniques (not uncommonly even from other fields) and no less important, intuition. Following this, the proposal is analysed, tried on a real system or its mathematical model, findings and errors are analysed again, the design is modified (or rejected) and the process continues until a satisfactory solution is found.

EAs work basically on the same principle, although, obviously, using less analytical analysis but more trial-and-error approach. It was found, however, that the process of selecting the most suitable solutions at each stage and producing the next iteration variants is, overall, largely intelligent and heuristic. EAs are capable to answer not only the question *how to do* something (how to control, in particular), but also the question *what to do* in order to meet the objective. It is therefore appealing to apply such a promising automated technique to a problem with no general solution at hand.

The guidance and flight control is not a totally unstudied area where no convincing guesses can be made and where no parallels with the existing solutions are possible. This fact allows to watch, understand and guide, to a certain extent, the process of evolution. It also enables to optimise the EA for the purposes of control design.

The latter is especially useful because there are still very little research done on artificial evolution of structures of controllers in particular. An overwhelming majority of EA applications is concerned with numeric optimisation. A few proponents of a more advanced use (e.g. Koza et al., 2000, De Jong & Spears, 1993) are keen to show the real scope of possible applications, including controller design.

1.2. Unmanned Aerial Vehicles And Shipboard Recovery

Over the past two decades, the use of UAVs is becoming a well accepted technique not only for the military applications but also in the civilian arena. Typical applications of UAVs range from such traditional military missions as battlefield surveillance, reconnaissance and target acquisition to atmospheric research, weather observation, coastal and maritime surveillance, agricultural and geological surveying, telecommunication signals retranslation, and search and rescue missions.

The critical parts of a UAV mission are the launch and recovery phases. Although some UAVs can be conventionally operated from runways, the ability of UAVs to be operated from confined areas, such as remote land sites, ships and oil rigs, greatly increase their practical applications. Such operations generally require the aircraft to either have Vertical Take- Off and Landing (VTOL) capability or some form of launch and recovery assistance.

Unlike launch, the ways of UAV recovery are numerous. Probably the most widely used method, apart from runway landing, is the parachute assisted recovery. Unfortunately, parachute recovery can hardly be used when the landing area is extremely limited (for example, a ship's deck) and in the presence of high winds and strong gusts.

The first practicable and widely used solution for shipboard recovery of a fixed-wing UAV was capturing by an elastic net. This method has been employed for the USN *RQ-2 Pioneer* UAV, first deployed in 1986 aboard the battleship USS *Iowa*. The recovery net is usually stretched above the stern of the ship and the aircraft is flown directly into the net. A serious disadvantage of this method is that it is quite stressful for the aircraft. Nevertheless, due to simplicity of the recovery gear and reasonably simple guidance and flight control during the approach, this technique is still very popular for maritime operations.

Other methods include such techniques as deep stall and perched recovery and various forms of convertible airframes. However, these methods often imply very specific requirements to the UAV design and high complexity of control.

In this work a novel recovery method is proposed. This method, named *Cable Hook Recovery*, is intended to recover small to medium-size fixed-wing UAVs on frigate size vessels. It is expected to have greater operational capabilities than the *Recovery Net* technique, which is currently the most widely employed method of recovery for similar class of UAVs, potentially providing safe recovery even in very rough sea and allowing the choice of approach directions.

There are two distinct areas in recovery design: design of the recovery technique itself and development of a UAV controller that provides flight control and guidance of the vehicle in accordance with the requirements of this technique. The controller should provide autonomous guidance and control during the whole recovery process (or its airborne stage).

It should be noted that there exists a number of control design techniques applicable to the area of guidance and flight control. They all have different features and limitations, producing the controllers with different characteristics. It is expected that linear control techniques will not be sufficient for control of the aircraft through the whole recovery stage due to large atmospheric disturbances, ship motion and aircraft constraints.

Under these conditions when so many factors remain uncertain during the process of development, even the very approach to the control problem is unclear. It is desirable that the controller design methodology allow to produce an optimally suitable controller even when faced with such uncertainties and that could be done with application of EAs.

2. EVOLUTIONARY ALGORITHMS

Over the hundred years of aviation history, various linear control methods have been successfully used in the aerospace area due to their simplicity and analytical justifiability. Despite their natural limitations, linear control techniques still remain as one of the most accepted design practices. However, growing demands to the performance of aircraft, and on the other hand, a remarkable increase in available computation power over the past years have led to significant growth if the popularity of nonlinear control techniques.

A principal difficulty of many nonlinear control techniques, which potentially could deliver better performance, is the impossibility or extreme difficulty to predict theoretically the behaviour of a system under all possible circumstances. Therefore, it becomes a challenging task to verify and validate the designed controller under all real flight conditions. There is a need to develop a consistent nonlinear control design methodology that enables to produce a required controller for an arbitrary nonlinear system while assuring its robustness and performance across the whole operational envelope at the same time.

The Evolutionary Algorithms (EAs) is a group of such stochastic methods which combine such important characteristics as robustness, versatility and simplicity and, indeed, proved the success in many applications, such as neural network optimisation (McLean &. Matsuda, 1998), finance and time series analysis (Mahfoud & Mani, 1996), aerodynamics and aerodynamic shape optimisation (McFarland & Duisenberg, 1995), automatic evolution of computer software and, of course, control (Chipperfield & Flemming, 1996).

2.1. Introduction To Evolutionary Algorithms

Evolutionary algorithm is an umbrella term used to describe computer-based problem solving systems which employ computational models of evolutionary processes as the key elements in their design and implementation. All major elements found in natural evolution are present in EAs. They are:

- Population, which is a set of individuals (or members) being evolved;
- Genome, which is all the information about an individual encoded in some way;
- Environment, which is a set of problem-specific criteria according to which the fitness of each individual is judged;
- Selection, which is a process of selecting of the fittest individuals from the current population in the environment;
- Reproduction, which is a method of producing the offspring from the selected individuals;
- Genetic operators, such as mutation and recombination (crossover), which provide and control variability of the population.

The process of evolution takes a significant number of steps, or generations, until a desired level of fitness is reached. The 'outcome' should not be interpreted as if some particular species are expected to evolve. The evolution is not a purposive or directed process. It is expected that highly fit individuals will arise, however the concrete form of these individuals may be very different and even surprising in many real engineering tasks. None of the size, shape, complexity and other aspects of the solution are required to be specified in advance, and this is in fact one of the great advantages of the evolutionary approach. The problem of initial guess value rarely exists in EA applications, and the initial population is sampled at random.

The first dissertation to apply genetic algorithms to a pure problem of mathematical optimisation was Hollstien's work (Hollstien, 1971). However, it was not until 1975, when John Holland in his pioneering book (Holland, 1975) established a general framework for application of evolutionary approach to artificial systems, that practical EAs gained wide popularity. Until present time this work remains as the foundation of genetic algorithms and EAs in general.

Now let us consider the very basics of EAs. A typical pseudo code of an EA is as follows:

- Create a {usually random} population of individuals;
- Evaluate fitness of each individual of the population;
- until not done {certain fitness, number of generations etc.}, do
- Select the fittest individuals as 'parents' for new generation;
- Recombine the 'genes' of the selected parents;
- Mutate the mated population;
- Evaluate fitness of the new population;
- end loop.

It may be surprising how such a simple algorithm can produce a practical solution in many different applications.

Some operations in EAs can be either stochastic or deterministic; for example, selection may simply take the best half of the current population for reproduction, or select the individuals at random with some bias to the fittest members. Although the latter variant can sometimes select the individuals with very poorly fitness and thus may even lead to temporary deterioration of the population's fitness, it often gives better overall results.

2.2. Evolutionary Algorithm Inside

There are several branches of EAs which focus on different aspects of evolution and have slightly different approaches to ongoing parameters control and genome representation.

In this work, a mix of different evolutionary methods is used, combining their advantages. These methods have the common names: Genetic Algorithms (GA), Evolutionary Programming (EP), Evolutionary Strategies (ES) and Genetic Programming (GP).

As noted above, all the evolutionary methods share many properties and methodological approaches.

2.2.1. Fitness Evaluation

Fitness, or objective value, of a population member is a degree of 'goodness' of that member in the problem space. As such, fitness evaluation is highly problem dependent. It is implemented in a function called fitness function, or more traditionally for optimisation methods, objective function. The plot of fitness function in the problem space is known as fitness landscape.

The word 'fitness' implies that greater values ('higher fitness') represent better solutions. However, mathematically this does not need to be so, and by optimisation both maximisation and minimisation are understood.

Although EAs are proved themselves as robust methods, their performance depends on the shape of the fitness landscape. If possible, fitness function should be designed so that it exhibits a gradual increase towards the maximum value (or decrease towards the minimum). In the case of GAs, this allows the algorithm to make use of highly fit building blocks, and in the case of ESs—to develop an effective search strategy quickly.

In solving real world problems, the fitness function may be affected by noise that comes from disturbances of a different nature. Moreover, it may be unsteady, that is, changing over time. Generally,

EAs can cope very well with such types of problem, although their performance may be affected.

It is accepted that the performance of an optimisation algorithm is measured in terms of objective function evaluations, because in most practical tasks objective evaluation takes considerably more computational resources than the algorithm framework itself. For example, in optimisation of the aerodynamic shape of a body, each fitness evaluation may involve a computer fluid flow simulation which can last for hours. Nevertheless, even for simple mathematical objective functions EAs are always computationally intensive (which may be considered as the price for robustness).

The computation performance may be greatly improved by parallelisation of the fitness evaluation. EAs process multiple solutions in parallel, therefore they are extremely easily adopted to parallel computing.

Another useful consequence of maintaining a population of solutions is the natural ability of EAs to handle multi-objective problems. A common approach – used in this work - to reduce a multiobjective tasks to a single-objective one, by summing up all the criteria with appropriate weighting coefficients, is not always possible. Unlike single point optimisation techniques, EAs can evolve a set of Pareto optimal solutions simultaneously. A Pareto optimal solution is the solution that cannot be improved by any criterion without impairing it by at least one other criterion, which is a very interesting problem on its own.

2.2.2. Genome Representation

In EA theory, much as well as in natural genetics, genome is the entire set of specifically encoded information that fully defines a particular individual. This section focuses only on numeric genome representation. However, EAs are not limited to numeric optimisations, and for more 'creative' design tasks a more sophisticated, possibly hierarchical, genome encoding is often required. In fact, virtually any imaginable data structure may be used as a genome. This type of problem is addressed in Section 2.5 Genetic Programming.

A commonly used genome representation in GAs is a fixed length binary string, called chromosome, with real numbers mapped into integers due to convenience of applying genetic operators, recombination (crossover) and mutation.

Accuracy is a common drawback of digital (discrete) machines, and care should be taken when choosing appropriate representation. For example, if *8* bit numbers are used and the problem determines the range of *[–1.0; 1.0]*, this range will be mapped into integers *[00000000; 11111111]* (*[0; 255]* decimal)1. As *255* corresponds to *1.0*, the next possible integer, *254*, will correspond to *0.99215686*, and there is no means of specifying any number between *0.99215686* and *1.0*.

The required accuracy is often set relative to the value, not the range. In this case, linear mapping may give too low accuracy near zero values and too high accuracy towards the end of the range. This was the reason for inventing the so called floating point number representation, which encodes the number in two parts: mantissa, which has a fixed range, and an integer exponent, which contain the order of the number. This representation is implemented in nearly all software and many hardware platforms which work with real numbers.

Another type of genome encoding which should be mentioned is the permutation encoding. It is used mostly in ordering problems, such as the classic Travelling Salesman Problem, where the optimal order of the given tokens is sought after. In this case, a chromosome can represent one of the possible permutations of the tokens (or rather their codes), for example, *'0123456789'* or *'2687493105'* for ten items. When this type of encoding is used, special care should be taken when implementing genetic operators, because 'traditional' crossover and mutation will produce mostly incorrect solutions (with some items included twice and some items missing).

Finally, a vector of floating point real values can be used as a chromosome to represent the problem that deals with real values. However, domain constraints handling should be implemented in most cases, as the floating point numbers, unlike integers, have virtually no highest and lowest values (in a practical sense). Moreover, special recombination and mutation operations should be used in this case.

2.2.3 Selection

Selection is the key element of all evolutionary algorithms. During selection, a generally fittest part of the population is chosen and this part (referred as mating pool) is then used to produce the offspring.

The requirements for the selection process are somewhat controversial. On the one hand, selection should choose 'the best of the best' to increase convergence speed. On the other hand, there should be some level of diversity in the population in order to allow the population to develop and to avoid premature convergence to a local optimum. This means that even not very well performing individuals should be included in the mating pool.

This question is known as the conflict between exploitation and exploration. With very little genetic diversity in a population, new areas in the search space become unreachable and the process stagnates. Although exploration takes valuable computation resources and may give negative results (in terms of locating other optima), it is the only way of gaining some confidence that the global optimum is found (Holland, 1975).

It should be noted that the optimal balance between exploitation and exploration is problem dependent. For example, real-time systems may want a quick convergence to an acceptable sub-optimal solution, thus employing strong exploitation; while engineering design which uses EAs as a tool is often interested in locating various solutions across the search space, or may want to locate exactly the global optimum. For the latter tasks, greater exploration and thus slower convergence is preferred.

Balance between exploitation and exploration can be controlled in different ways. For example, intuitively, stronger mutation favours greater exploration. However, it is selection that controls the balance directly. This is done by managing the 'strength' of selection. Very strong selection realises exploitation strategy and thus fast convergence, while weak selection allows better exploration.

A general characteristic that describes the balance between 'perfectionism' and 'carelessness' of selection is known as selection pressure or the degree to which the better individuals are favoured. Selection pressure control in a GA can be implemented in different ways; a very demonstrative parameter is the size of the mating pool relative to the size of the population. As a rule, the smaller the mating pool, the higher the selection pressure.

A quantitative estimation of the selection pressure may be given by the take-over time (Goldberg & Deb, 1991). With no mutation and recombination, this is essentially the number of generations taken for the best member in the initial generation to completely dominate the population.

The efficiency of one or another selection method used in EAs largely depends on population properties and characteristics of the whole algorithm. A theoretical comparison of the selection schemes may be found in (Goldberg & Deb, 1991).

First, all selection methods can be divided into two groups: stochastic and deterministic. Deterministic methods use the fitness value of a member directly for selection. For example, the best half of the population may be selected or all individuals with the fitness better than a given value may be included in the mating pool. In contrast, stochastic selection methods use fitness only as a guide, giving the members with better fitness more chances to be selected allowing greater exploration. However, deterministic methods can also be tuned to allow greater exploration. For example, every second member in a sorted by fitness list can be taken for reproduction instead of the best half of the population.

One of the simple ways to reduce the possible impact of stochastic sampling errors is to guarantee that the best, or elite, member(s) is always selected for reproduction. This approach is known as Elitism. In effect, elitism is the introduction of a portion of deterministic selection into a stochastic selection procedure. In most cases, elitism assumes that the elite member is not only selected, but also propagated directly to the new population without being disrupted by recombination or mutation. This approach ensures that the best-so-far achievement is preserved and the evolution does not deteriorate, even temporarily.

Another feature which may be applied to any selection scheme is population overlapping. The fraction of the old population which is replaced with the fresh members is called a generation gap (De Jong, 1975). Nothing particularly advantageous is found in overlapping schemes, although they may be useful for some problems, in particular for steady and noiseless environments (Goldberg & Deb, 1991).

It should be also noted that some reproduction schemes allow multiple selection of one member, while others do not. The former case (also referred to as replacement) means that the selected member is returned back to the mating pool and thus may be selected again in the same generation. This feature is often used in stochastic selection methods such as fitness proportional selection and tournament selection.

2.2.3.1 Deterministic Selection

All members are straightforwardly sorted according to their fitness value and some of the best members are picked up for reproduction. A certain number of members (or population percentage) is usually chosen, or the members may be selected one by one and reproduced until the next generation population is filled up.

As noted before, deterministic methods are more suitable for the algorithms with small populations (less than about *20–40* members). They are therefore used in the areas where small populations are desirable (e.g. when fitness evaluation is extremely costly) or where it is traditionally adopted (in particular in Evolutionary Strategies, see Section 2.4).

2.2.3.2 Fitness Proportional Selection And Fitness Scaling

In fitness proportional selection, all individuals receive the chances to reproduce that are proportional to their objective value (fitness). There are several implementations of this general scheme which vary in stochastic properties and time complexity: roulette wheel (Monte Carlo) selection, stochastic remainder selection and stochastic universal selection. The roulette wheel method is described here in more detail.

The analogy with a roulette wheel arises because one can imagine the whole population forming a roulette wheel with the size of any individual's slot proportional to its fitness. The wheel is then spun and the 'ball' thrown in. The probability of the 'ball' coming to rest in any particular slot is proportional to the arc of the slot and thus to the fitness of the corresponding individual (Coley, 1999).

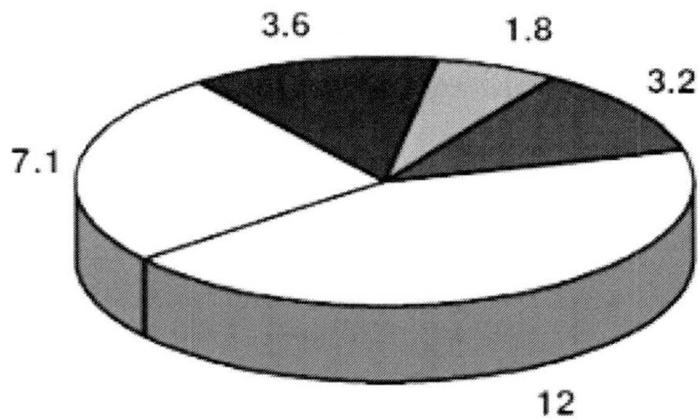

Figure 1. Roulette wheel selection

There are no means to control the selection pressure and the convergence speed: they are determined entirely by the fitness of each individual.

However, such a control is often necessary. If for example, a fit individual is produced, fitness proportional selection with replacement can allow a large number of copies of this individual to flood the subsequent generations.

One of the methods intended to overcome this problem and to maintain a steady selection pressure is linear fitness scaling (Coley, 1999). Linear fitness scaling works by pivoting the fitness of each individual about the average population fitness. The scale is chosen so that an approximately constant proportion of copies of the best members is selected compared to the 'average member'.

There are some more sophisticated scaling techniques, such as sigma scaling (Coley, 1999), in which the (expected) number of trials each member receives is adjusted according to the standard deviation of the population fitness.

2.2.3.3 Ranking Selection

This is a development of the fitness proportional selection, aimed to achieve greater adaptability and to reduce stochastic. The idea represents the combination of fitness proportional and deterministic selection. The population is sorted according to the fitness, and a rank is assigned to each individual. After assigning the rank, a proportionate selection is applied as described in the previous section, using rank values instead of fitness.

Ranking has two main advantages before fitness proportional selection (even that with fitness scaling). First, the required selection pressure can be controlled more flexibly by applying a specific rank assignment function. Second, it softens stochastic errors of the search, which can be especially destructive for the fitness functions affected by noise. If a particularly fit member is generated that stands well off the whole population. Even if the proportionate selection is constrained by fitness scaling, this best member will be greatly favoured, whilst the rest of the population will receive very low selection pressure because the differences between their fitness values are insignificant as compared to the 'outlier'. In contrast, ranking will establish a predefined difference between the neighbouring members, ensuring an adequate selection pressure for the whole population.

2.2.3.4 Tournament Selection

Tournament selection is a simple yet flexible stochastic selection scheme. Choose some number s of individuals randomly from a population and then select the best individual from this group. Repeat as many times as necessary to fill up the mating pool. This somewhat resembles the tournaments held between *s* competitors. As a result, the mating pool, being comprised of tournament winners, has a higher average fitness than the average population fitness.

The selection pressure can be controlled simply by choosing appropriate tournament size *s*. Obviously, the winner from a larger tournament will, on average, have a higher fitness than the winner of a smaller tournament. In addition, selection pressure can be further adjusted via randomisation of the tournament.

2.2.4. Recombination

Recombination allows solutions to exchange the information in a way similar to that used by a biological organism undergoing sexual reproduction. This effective mechanism allows to combine parts of the solution (building blocks) successfully found by parents. Combined with selection, this scheme produces, on average, fitter offspring. Of course, being a stochastic operation, recombination can produce 'disadvantaged' individuals as well; however, they will be quickly perished under selection.

Recombination is usually applied probabilistically with a certain probability. For GAs, the typical value is between *0.6* and *0.8*; however, the values up to *1.0* are common.

2.2.4.1. Alphabet Based Chromosome Recombination

In essence, recombination is 'blending' the information of two (or more) genomes in some way. For typical GAs, an approach from natural genetics is borrowed. It is known as crossover. During crossover, chromosomes exchange equal parts of themselves. In its simplest form known as single-point crossover, two parents are taken from the mating pool. A random position on the chromosome (locus) is chosen. Then, the end pieces of the chromosomes, starting from the chosen locus, are swapped.

Single-point crossover can be generalised to *k*-point crossover, when *k* different loci are chosen and then every second piece is swapped. However, according to De Jong's studies (De Jong, 1975) and also (Spears & Anand, 1991), multi-point crossover degrades overall algorithm performance increasingly with an increased number of cross points.

There is another way of swapping pieces of chromosomes, known as uniform crossover. This method does not select crossover points. Instead, it considers each bit position of the two parents one by one and swaps the corresponding bits with a probability of *50%*. Although the uniform crossover is, in a sense, an extreme case of multi-point crossover and can be considered as the most disruptive its variant, both theoretical and practical results (Spears & Anand, 1991) show that uniform crossover outperforms k-point crossover in most cases.

2.2.4.2 Real Value Recombination

Real-coded EAs require a special recombination operator. Unlike bit strings, real parameters are not deemed as strings that can be cut into pieces. Instead, they are processed as a whole in a common mathematical way. Due to rather historical reasons, real-coded EAs were mostly developing under the influence of Evolutionary Strategies and Evolutionary Programming (see Section 2.4). As a result, real-value recombination has not been properly considered until the fairly recent past ('90s). Nevertheless, a number of various recombination techniques have been developed. Detailed analysis of them is available in (Beyer & Deb, 2001, Deb et al., 2001 and Herrera et al., 1998).

The simplest real-value recombination one can think of is the averaging of several parents, which is known as arithmetic crossover. This method produces one offspring from two or more parents. Averaging may be weighted according to parents' fitness or using random weighting coefficients.

Self-adaptive recombination create offspring statistically located in proportion to the difference of the parents in the search space. These recombination operators generate one or two children according to a probability distribution over two or more parent solutions where if the difference between the parent solutions is small, the difference between the child and parent solutions should also be small.

The most popular approach is to use a uniform probability distribution—the so called 'blend crossover', *BLX*. The *BLX* operator randomly picks a solution in the range

$$\left[x_1 - \alpha\left(x_2 - x_1\right); x_2 + \alpha\left(x_2 - x_1\right)\right] \tag{1}$$

for two parents $x_1 < x_2$. is the parameter which controls the spread of the offspring interval beyond the range of the parents' interval [x_1; x_2].

Other approaches suggest non-uniform probability distribution. The Simulated Binary Crossover (*SBX*) uses a bimodal probability distribution with its mode at the parent solutions. It produces two children from two parents. This technique has been developed by K. Deb and his students in 1995. As the name suggests, the *SBX* operator simulates the working principle of the single-point crossover operator on binary strings.

A different approach is demonstrated by the Unimodal Normal Distribution Crossover (*UNDX*) (Ono & Kobayashi, 1997). It uses multiple (usually three) parents and create offspring solutions around the centre of mass of these parents. *UNDX* uses a normal probability distribution, thus assigning a small probability to solutions away from the centre of mass. Another mean-centric recombination operator is the Simplex Crossover (*SPX*). It differs from *UNDX* by assigning a uniform probability distribution for creating offspring in a restricted region marked by the parents.

Although both mean-centric and parent-centric recombination methods were found to exhibit self-adaptive behaviour for real-coded GAs similar to that of ESs (see Section 2.4), in a number of reports parent-centric methods were found generally superior (Deb et al., 2001).

2.2.5. Mutation

Mutation is another genetic operator borrowed from nature. However, unlike recombination, which is aimed at producing better offspring, mutation is used to maintain genetic diversity in the population from one generation to the next in a explorative way.

Not unlike recombination, mutation works differently for alphabet-based chromosomes and real-coded algorithms. However, in both cases it is merely a blind variation of a given individual.

2.2.5.1. Bit String Mutation

In nearly all ordinary GAs, mutation is implemented as variation of a random bit in the chromosome. Each bit in the chromosome is considered one by one and changed with certain probability P_m. Bit change can be applied either as flipping (inverting) of the bit or replacing it with a random value. In the latter case, the actual mutation rate will be twice as low, because, on average, half of the bits are replaced with the same value.

In GAs, mutation is usually controlled through mutation probability P_m. As a rule, Gas put more stress on recombination rather than on mutation, therefore typical mutation rates are very low, of the order of*0.03* and less (per bit). Rates close to *0.001* are common. Nevertheless, mutation is very important because it prevents the loss of building blocks which cannot be recovered by recombination alone.

Mutation probability can be supressed in a number of ways. However, this may easily have an adverse effect—mutation is known for the ability to 'push' the stagnated process at the later stages.

Another technique to avoid stagnation is so called 'hypermutation'. Hypermutation is a method in which mutation probability is significantly (*10–100* times) increased for a small number of generations (usually one) during the run, or such a high rate is applied constantly to a certain part of the population (e.g. *20%*). Both hypermutation and random immigrants techniques are especially effective for dynamic environments, where the fitness landscape can change significantly over time (Grefenstette, 1999).

The above mutation control methods have a common drawback: they are not selective. Sometimes an individual approach may be desirable. In particular, stronger mutation might be applied to the weakest members, while less disruptive mutation to the best individuals. Alternatively, some more sophisticated methods can be developed. Such fine tuning is a more common feature of Evolutionary Strategies (see section 2.4) and real-coded GAs.

Some more direct control methods utilise additional bits in the genotype which do not affect the phenotype directly. However, these bits control the mutation itself. In the simplest case, a single flag bit can control the applicability of mutation to the respective parameter when zero value disables mutation and unity enables it.

2.2.5.2. Real Value Mutation

Implementation of a real-value mutation is rather more straightforward than that of alphabet strings. Unlike the latter, it is not an operation directly inspired by nature; however, as the real-coded algorithms generally do not use tricky encoding schemes and have the same problem-space and genotype values of the parameters, real-value mutation can be considered as the operation working on a higher level, up to direct phenotype mutation for function optimization problems.

Mutation to a real value is made simply by adding a random number to it. It is evident that the strength of real-value mutation can be controlled in a very convenient way, through the variance of the distribution (or the window size for the uniform distribution). Like with the common GAs that operate string-type chromosomes, mutation strength can be adapted using many different techniques, from simple predefined linear decrease to sophisticated adaptation strategies

2.2.6. Measuring Performance

Performance of an EA is the measure how effective the algorithm is in search of the optimal solution. As evolutionary search is a stochastic and dynamic process, it can hardly be positively measured by a single figure. A better indicator of the performance is a convergence graph, that is, the graph 'fitness vs. computation steps'.

This figure presents two typical convergence graphs of two independent runs of the same GA with different sets of parameters. It can be seen that although the first run converges faster, it stagnates too early and does not deliver the optimal solution. Therefore, not the convergence speed nor time to reach a specified value, nor any other single parameter can be considered as a sole performance measure.

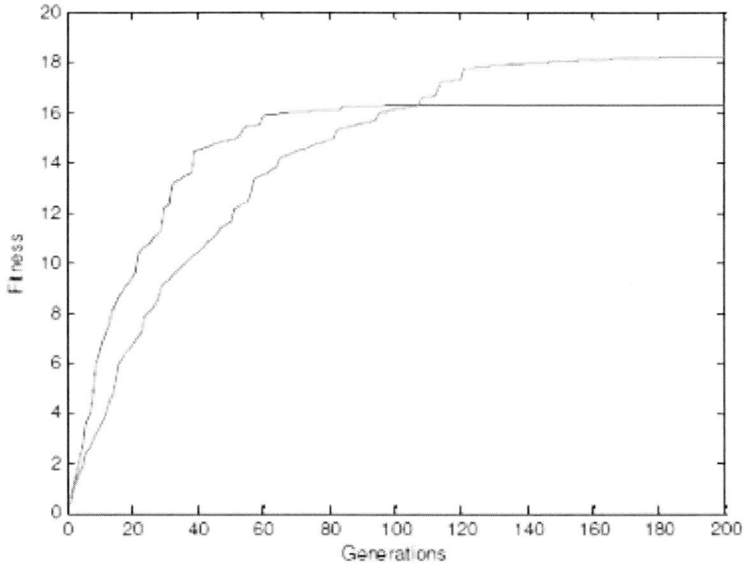

Figure 2. Typical convergence graphs

De Jong in (De Jong, 1975) used two measures of the progress of the GA: the off-line performance and the on-line performance. The off-line performance is represented by the running average of the fitness of the best individual, f_{max}, in each population:

$$f_{off}(g) = \frac{1}{g} \sum_{i=1}^{g} f_{max}(g) \tag{2}$$

where g is the generation number. In contrast, the on-line performance is the average of all fitness values calculated so far:

$$f_{on}(g) = \frac{1}{g}\sum_{i=1}^{g} f_{avg}(g) \tag{3}$$

The on-line performance includes both good and bad guesses and therefore reflects not only the progress towards the optimal solution, but also the resources taken for such progress.

Another useful measure is the convergence velocity:

$$V(g) = \ln\sqrt{\frac{f_{max}(g)}{f_{on}(1)}} \tag{4}$$

In the case of a non-stationary dynamic environment, the value of previously found solutions is irrelevant at the later steps. Hence, a better measure of optimisation in this case is the current-best metric instead of running averages.

When comparing the performance of different algorithms, it is better to use the number of fitness evaluations instead of the number of generations as the argument for performance characteristics.

2.2.7. The Problem Of Convergence And Genetic Drift

Usually, the progress of EAs is fast at first and then loses its speed and finally stagnates. This is a common scenario for optimisation techniques. However, progressing too quickly due to greedy exploitation of good solutions may result in convergence to a local optimum an thus in low robustness. Several methods that help to control the convergence have been described in the above sections, including selection pressure control and adaptation of recombination and mutation rates.

However, it is unclear to which degree and how in particular to manage the algorithm's parameters. What is 'too fast' convergence? Unfortunately, the current state of EA theory is inadequate to answer this question *before* the EA is initiated.

For the most of the real-world problems, it takes several trial runs to obtain an adequate set of parameters. The picture of convergence process, as noted before, is not a good indicator of the algorithm's characteristics. In contrast, the plot of genetic diversity of the population against generation number (or fitness function evaluations) gives a picture which can explain some performance problems. If the population quickly loses the genetic diversity, this usually means a too high initial selection pressure. The further saturation may be attributed to reduction of selection pressure at the later stages. The loss of genetic diversity is known as genetic drift (Coley, 1998).

2.2.8. Schema Theory

Schema theory has been developed by John Holland (Holland, 1975) and popularised by David Goldberg (Goldberg, 1989) to explain the power of binary-coded genetic algorithms. More recently, it has been extended to real-value genome representations (Eshelman & Schaffer, 1993) and tree-like *S*-expression representations used in genetic programming (Langdon & Poli, 2002, O'Reilly & Oppacher, 1995). Due to the importance of the theory for understanding GA internals, it is mentioned here, though it is not within the scope of this work to discuss it in details.

2.3. Genetic Algorithms

Genetic algorithms are one of the first evolutionary methods successfully used to solve practical problems, and until now they remain one of the most widely used EAs in the engineering field. John Holland in (Holland, 1975) provided a general framework for GAs and a basic theoretical background, much of which has been discussed in the former sections. There are more recent publications on the basics of GA, for example (Goldberg, 1989).

The basic algorithm is exactly as in the Section 2.1; however, several variations to this scheme are known. For example, Koza (Koza, 1992) uses separate threads for asexual reproduction, crossover and mutation, chosen at random; therefore, only one of these genetic operators is applied to an individual in each loop, while classical GA applies mutation after crossover independently.

One of the particularities of typical GAs is genome representation. The vast majority of GAs use alphabet-based string-like chromosomes described in Section 2.2.2, although real coded GAs are gaining wider popularity. Therefore, a suitable mapping from actual problem space parameters to such strings must be designed before a genetic search can be conducted.

The objective function can also have the deceptive properties as in most practical cases little is known about the fitness landscape. Nevertheless, if the fitness function is to be designed for a specific engineering task (for example, an estimate of the flight control quality, as will be used in this study later), attention should be paid to avoiding rugged and other GA-difficult fitness landscapes.

Of the two genetic operators, recombination (crossover) plays the most important role in Gas. Typical probability of crossover is *0.6* to *0.8* and even up to *1.0* in some applications. On the contrary, mutation is considered as an auxiliary operator, only to ensure that the variability of the population is preserved. Mutation probabilities range from about *0.001* to *0.03* per bit.

Population size is highly problem dependent; however, typical GAs deal with fairly large or at least moderate population sizes, of the order of *50* to *300* members, although smaller and much larger sizes (up to several thousands) could be used.

Although by far the largest application of GAs is optimisations of different sorts, from simple function optimisations to multi-parameter aerodynamic shape optimisation (McFarland & Duisenberg, 1995) and optimal control (Chipperfield & Flemming, 1996), GAs are suitable for many more tasks where great adaptation ability is required, for example, neural networks learning (Sendhoff & Kreuz, 1999) and time series prediction (Mahfoud & Mani, 1996). The potential of GA application is limited virtually only by the ability to develop a suitable encoding.

2.4. Evolutionary Strategies And Evolutionary Programming

Evolutionary Programming was one of the very first evolutionary methods. It was introduced by Lawrence J. Fogel in the early 1960s (Fofel, 1962), and the publication (Fogel et al., 1966) by Fogel, Owens and Walsh became a landmark for EP applications. Originally, EP was offered as an attempt to create artificial intelligence. It was accepted that prediction is a keystone to intelligent behaviour, and in (Fogel et al., 1966) EP was used to evolve finite state automata that predict symbol strings generated from Markov processes and non-stationary time series.

In contrast, Evolutionary Strategies appeared on the scene in an attempt to solve a practical engineering task. In 1963, Ingo Rechenberg and Hans-Paul Schwefel were conducting a series of wind tunnel experiments in Technical University of Berlin trying to optimise aerodynamic shape of a body. This was a laborious intuitive task and the students tried to work strategically. However, simple gradient and coordinate strategies have proven to be unsuccessful, and Rechenberg suggested to try random changes in the parameters defining the shape, following the example of natural mutations and selection.

As it can be seen, both methods are focusing on behavioural linkage between parents and the offspring rather than seeking to emulate specific genetic operators as observed in nature. In addition, unlike GAs, natural real-value representation is predominantly used. In the present state, EP and ES are very similar, despite their independent development over 30 years, and the historical associations to finite state machines or engineering field are no longer valid. In this study, ES approach is employed, so further in this section Evolutionary Strategies are described, with special notes when the EP practice is different.

2.4.1. Self-Adaptation

One of the most important mechanisms that differs ES from the common GAs is endogenous control on genetic operators (primarily mutation). Mutation is usually performed on real-value parameters by adding zero mean normally distributed random values. The variance of these values is called step size in ES.

The adaptation of step size rules can be divided into two groups: pre-programmed rules and adaptive, or evolved, rules. The pre-programmed rules express a heuristic discovered through extensive experimentation. One of the earliest examples of pre-programmed adaptation is Rechenberg's (1973)*1/5* rule. The rule states that the ratio of successful mutations to all mutations should be *1/5* measured over a number of generations. The mutation variance (step size) should increase if the ratio is above*1/5*, decrease if it is below and remain constant otherwise. The variance is updated every k generations according to:

$$\sigma^{(g)} = \begin{cases} \sigma^{(g-k)} / c & n_s > 1/5 \\ \sigma^{(g-k)} \cdot c & n_s < 1/5 \\ \sigma^{(g-k)} & n_s = 1/5 \end{cases} \quad (5)$$

where n_s is the ratio of successful mutations and *0.817* $c < 1$ is the adaptation rate. The lower bound $c = 0.817$ has been theoretically derived by Schwefel for the sphere problem (Ursem, 2003). The upper index in parenthesis henceforth denotes the generation number.

The other approach is the self-adaptive (evolved) control where Schwefel (Schwefel, 1981) proposed to incorporate the parameters that control mutation into the genome. This way, an individual $a = (x,\sigma\sigma)$ consists of *object variables* (sometimes referred as *describing parameters*) x and *strategy parameters*$\sigma\sigma$. The strategy parameters undergo basically the same evolution as object variables: they are mutated and then selected together, though only on the basis of objective performance, on which strategy parameters have indirect influence. The underlying hypothesis in this scheme is that good solutions carry good strategy parameters; hence, evolution discovers good parameters while solving the problem.

2.5. Genetic Programming

Genetic programming (GP) is an evolutionary machine learning technique. It uses the same paradigm as genetic algorithms and is, in fact, a generalisation of GA approach. GP increases the complexity of the structures undergoing evolution. In GP, these structures represent hierarchical computer programs of varying size and shape.

GP is a fairly recent EA method compared to other techniques discussed before in this chapter. The first experiments with GP were reported by Stephen Smith (Smith, 1980) and Nichael Cramer (Gramer, 1985). However, the first seminal book to introduce GP as a solid and practical technique is John Koza's 'Genetic Programming', dated 1992.

In GP, each individual in a population is a program which is executed in order to obtain its fitness. Thus, the situation is somewhat opposite to GAs: the individual is a 'black box' with an arbitrary input and some output. The fitness value (often referred to as *fitness measure* in GP) is usually obtained through comparison of the program's output with the desired output for several input test values (*fitness cases*). However, fitness evaluation in GP is problem dependent and may be carried out in a number of ways. For example, the fitness of a program controlling a robotic animal may be calculated as the number of food pieces collected by the animal minus resources taken for search (e.g. path length). When seeking a function to fit the experimental data, the deviation will be the measure of fitness.

One of the characteristics of GP is enormous size of the search space. GP search in the space of possible computer programs, each of which is composed of varying number of functions and terminals. It can be seen that the search space is virtually incomprehensible, so that even generation of the initial random population may represent some difficulties. Due to that, GP typically works with very large populations of hundreds and even thousands of members. Two-parent crossover is usually employed as the main genetic operator, while mutation has only a marginal role or is not used at all.

2.5.1. Genome Representation In Gp And S-Expressions

Unlike linear chromosomes in GAs, genomes in GP represent hierarchical, tree-like structures. Any computer program or mathematical expression can be depicted as a tree structure with functions as nodes and terminals as leaves. For example, let us consider an expression for one of the roots of a square equation $ax^2 + bx + c = 0$:

$$\frac{-b+\sqrt{b^2-4ac}}{2a}$$

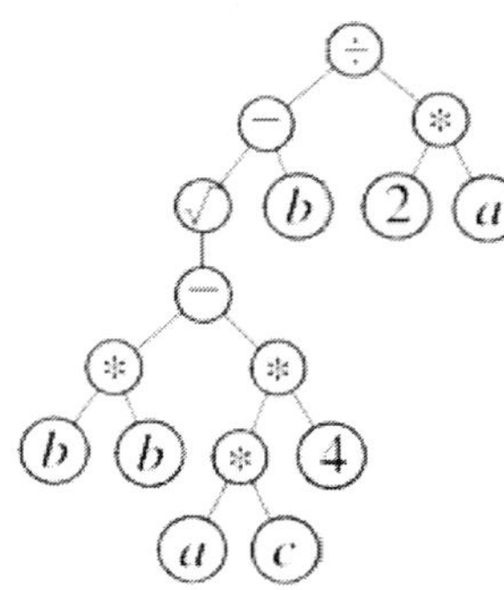

Figure 3. Tree-like representation of an expression

The convenience of such tree-like structures is that they can be easily modified on the sub-tree level. Any sub-tree can be taken out and replaced with another one, preserving syntax validity of the expression.

However, the trees such as shown in Fig. 3 should be encoded in some computer-readable form for actual GP implementation. This can be done in a number of ways. Some systems (e.g. MATLAB) provide built-in mechanisms for storage and operation on hierarchical structures. If this is not available, string representations are employed. An expression or a program can be encoded in a common for imperative languages way; for example, the formula for the root of a square equation from Fig. 3 can be written as

$$(sqrt(b*b-4*a*c)-b)/(2*a) \tag{6}$$

Unfortunately, such representation, although being mathematically readable, is inconvenient to handle in a GP way. It has to be parsed to the tree-like form for every operation. Therefore, another syntax is traditionally used.

One can note that in the trees such as the ones above, an operation always precedes its arguments on the branch, e.g. instead of '$a + b$' it reads '$+ a\ b$'. This notation is known as prefix notation or Polish notation. It is used in the programming language LISP and its derivatives— certainly not the most human-friendly language but very flexible and useful in many areas, and is one of the most popular languages in GP field.

There are extensions to the tree-based GP. Most of them employ decomposition of the programs into sub-trees (modules) and evolving these modules separately. One of the most widely used methods of this kind is Koza's Automatically Defined Functions (ADF) (Koza, 1994). In ADF approach, the program is split into a main tree and one or more separate trees which take arguments and can be called by the main program or each other. In another approach, code fragments from successful program trees are automatically extracted and are held in a library, and then can be reused in the following generations by any individual via library calls.

However, tree-based GP is not the only option. It is possible to express a program as a linear sequence of commands. One of the examples of linear GP systems is stack-based GP (Perkins, 1994). In stack-based languages (such as Forth) each program instruction takes its arguments from a stack, performs its calculations and then pushes the result back onto the stack. For example, the sequence *1 2* +. pushes the constants *1* and *2* onto the stack, then '+' takes these values from the stack, performs the addition and pushes the result 3 back. The final dot extracts and prints out the result. The notation such as '*1 2+*' is the opposite to that used in LISP and is called *reverse Polish notation* (or *postfix notation*).

2.5.1.1. Function Set And Terminal Set

When designing a GP implementation, proper care should be taken for choosing the function and terminal sets. The function set F = $\{f_1, f_2,...,f_{nf}\}$ is the set of functions from which all the programs are built. Likewise, the terminal set $T = \{a_1, a_2,...,a_{nt}\}$ consists of the variables available for functions. In principle, the terminals can be considered as functions with zero arguments and both the sets can be combined in one set of primitives C = F U T.

The choice of an appropriate set of functions and variables is crucial for successful solution of a particular problem. Of course, this task is highly problem dependent and requires significant insight. In some cases, it is known in advance that a certain set is sufficient to express the solution to the problem at hand.

However, in most practical real-world problems the sufficient set of functions and terminals is unknown. In these cases, usually all or most of the available data is supplied to the algorithm or iterative design is employed when additional data and functions are added if the current solution is unsatisfactory. As a result, the set of primitives is often far from the minimal sufficient set.

The effect of adding extraneous functions is complex. On the one hand, an excessive number of primitives may degrade performance of the algorithm, similar to choosing excessive genome length in GA. On the other hand, a particular additional function or variable may dramatically improve performance of both the algorithm and solution for a particular problem. For example, addition of the integral of error $\int(H-Hset)dt$ as an input to altitude hold autopilot allows to eliminate static error and improve overall performance. Alternatively, the integrator function may be introduced along with both the current altitude reading H and the desired altitude H_{set}.

2.5.2 Initial Population

Generation of the initial population in GP is not as straightforward as it usually is in conventional GAs. It has been noted above that the shape of a tree has (statistically) an influence on its evolution and that both sparse and bushy trees should be presented in the initial population. To this end, a so called 'ramped half-and-half' method, suggested by Koza (Koza, 1992), is typically used in GP. In this method, half of the trees in the population are generated as full trees and another half as random trees.

The 'ramped half-and-half' method employs two techniques for random tree generation: the 'full' method and the 'grow' method. Both of them start from choosing one of the functions from the function set F at random. It becomes the root of the new tree. Then, for each of the inputs of this function, a new primitive is selected with uniform probability. If the path from the root to the current point is shorter than the specified maximum depth, the new primitive is selected from the function set F for the 'full' method and from the union set C for the 'grow' method. If the path length reaches the specified depth, a terminal from the set T is selected at random for both methods. The process continues until the tree is complete, i.e. all the inputs are connected.

2.5.3 Genetic Operators

2.5.3.1. Crossover

Crossover is usually the most important genetic operator in GP. Its classic variation (Koza, 1992) produces two children trees from two parent trees by exchanging randomly selected sub-trees of each parent. Both parents are selected using one of the stochastic selection methods such as fitness proportional selection and tournament selection. The crossover operation begins by choosing, using a uniform probability distribution, one random point in each parent independently to be the crossover point for that parent. The point may be a node as well as a leaf. Then, the sub-trees that have roots at the crossover points are removed from the parents, and the sub-tree from the second parent is inserted in place of the removed sub-tree of the first parent.

In terms of S-expressions, the sub-tree crossover is equivalent to exchanging the sublists of the parental lists. Considering the example from Fig. 3, the parental solutions are

$$\left(-(*\quad x\quad x)\quad (*(*\quad x\quad y)\,2)\right) \text{ and } \left(/\quad (+\quad (*\quad x\quad x)\quad y)\quad 2\right) \tag{7}$$

The sub-lists corresponding to the selected crossover fragments are emphasized. These sublists are swapped between the parents and the following offspring are produced:

$$\left(-(*\quad x\quad x)\quad \left(*(+\quad (*\quad x\quad x)\quad y)\,2\right)\right) \text{ and } \left(/\quad (*\quad x\quad y)\quad 2\right) \tag{8}$$

It can be noted that in such a simple operation, syntactic validity of the resulting expressions is always preserved.

To avoid excessive growth of the branches of the program trees, a maximum depth value is usually established. If the crossover operation produces an offspring of impermissible depth, this offspring is disregarded and its parent is reproduced as is. Koza (Koza, 1992) uses the default maximum depth of 17. Even such a modest value allows creation of the trees of enormous size, up to $2^{17} - 1 = 131071$ (for binary trees).

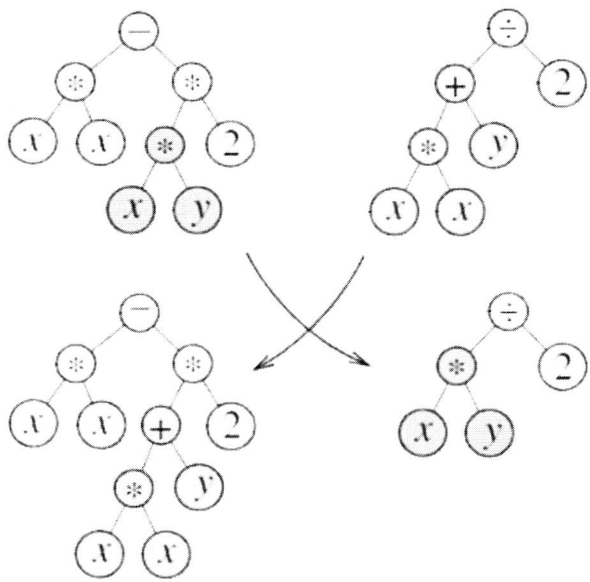

Figure 4. Sub-tree crossover: $x^2 - 2xy$ crossed with $(x^2 + y)/2$ to produce $x^2 - 2(x^2 + y)$ and $xy/2$

2.5.3.2. Other Genetic Operators

Reproduction as such is simply copying of the selected individual into the next generation population. In classical GP (Koza, 1992), about *10%* of the population is selected for simple reproduction and *90%* is reproduced through crossover.

Mutation is typically a replacement of a randomly selected sub-tree of an individual with a new randomly generated sub-tree. A special case is point mutation, when a single random terminal is inserted in place of a sub-tree. In fact, point mutations sometimes happen during crossover operation. In general, mutation is less needed for GP than for GAs, because crossover alone can reintroduce genetic diversity. In many practical applications, mutation is not used at all.

Permutation is changing of the order of arguments of a randomly selected function. The effect and usefulness of permutation is roughly the same as that of mutation.

Editing is changing the shape and structure of a tree while maintaining its semantic meaning. Usually this implies a mathematical simplification of the expression. For example, the sub-trees which can be immediately evaluated may be replaced with a corresponding terminal, e.g. the expression (+ *2 3*) may be replaced with (*5*).

Simplification of the solutions may reduce (sometimes significantly) their length and thus reduce the computation time needed for their evaluation. In addition, shorter expressions are less vulnerable to disruption caused by crossover and mutation as there are fewer chances that they will be torn apart.

2.5.4. Convergence In Gp

The question of convergence in GP is more complicated than in GAs. Generally, it is assumed in EA theory that the population converges when it contains substantially similar individuals (Langdon & Poli, 1992). Unlike conventional GAs, which have one-to-one mapping between genotype and phenotype, this rarely happens in GP. The search space in GP is essentially bigger than the phenotypic search space of the problem at hand. Any solution to the problem may be represented in an infinite number of ways.

Therefore, the population in GP may contain significantly different individuals (in terms of size and shape) and continue to evolve while yielding practically similar solutions to the problem.

However, the genotype cluster does not stabilise and continues to evolve. Since then, most of the highly fit individuals are produced by adding relatively insignificant branches to the successful core that came from the common ancestor. Therefore, each descendant genotype tends to be bigger than its parents. This results in a progressive increase in size known as bloat.

2.5.5. Bloat

The rapid growth of programs produced by GP is known since the beginning (Koza, 1992). As already noted, this growth need not to be correlated with increase of fitness because it consists of the code that does not change the semantics of the evolving programs. The rate of growth varies depending upon the particular GP paradigm being used, but usually exponential rates are observed (Nordin & Banzhaf, 1995).

For greater detail of different theories of bloat the reader is referred to (Langton, 1999) or (Langton & Poli, 2002). Generally, it should be noted that GP crossover by itself does not change the average program size. Bloat arises from the interaction of genetic operators and selection, i.e. selection pressure is required.

The most commonly (if not always) used restrictive technique is size and depth limits. Its implementation is already described in Section 2.5.3.1. It should be noted that the actual experiments (Langton & Poli, 2002) indicate that the populations are quickly affected by even apparently generous limits. Another commonly used approach is to give some preference to smaller solutions. The 'penalty' for excessive length may be included in fitness evaluation. However, in order not to degrade the performance of the algorithm, this component should be small enough so that it would have effect only for the solutions with identical phenotypic performance.

3. AIRCRAFT FLIGHT CONTROL

Not unlike the generic control approach, aircraft flight control is built around a feedback concept. Its basic scheme is shown in Fig. 5. The controller is fed by the difference between the commanded reference signal r and the system output y. It generates the system control inputs u according to one or another algorithm.

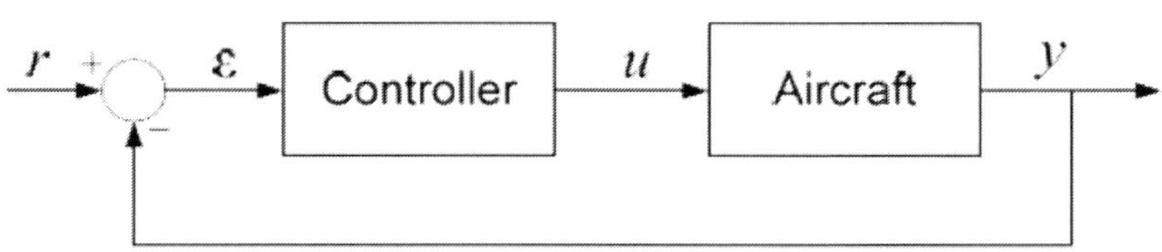

Figure 5. Feedback system (one degree of freedom)

One of the main tasks of flight control as an engineering discipline is design of the controllers which enable a given aircraft to complete a defined mission in the most optimal manner, where optimality is based on mission objective.

3.1. Intelligent Control

Intelligent control is a general and somewhat bold term that describes a diverse collection of relatively novel and non-traditional control techniques based on the soft computing approach. These include neural networks, fuzzy logic, adaptive control, genetic algorithms and several others. Often they are combined with each other as well as with more traditional methods.

Evolutionary and *genetic algorithms* (EAs, GAs) are global optimisation techniques applicable to a broad area of engineering problems. They can be used to optimise the parameters of various control systems, from simple PID controllers (Zein-Sabatto & Zheng, 1997) to fuzzy logic and neural network driven controllers (Bourmistrova, 2001, Kaise & Fujimoto, 1999). Another common design approach is evolutionary optimisation of trajectories, accompanied by a suitable tracking controller (e.g. (Wang & Zalzala, 1996)). An elaborated study of applications of EAs to control and system identification problems can be found in (Uzrem, 2003). As discussed above, Genetic Algorithms (in the form of*Genetic Programming*) are able to evolve not only the parameters, but also the structure of the controller.

In general, EAs require substantial computational power and thus are more suitable for offline optimisation. However, online evolutionary-based controllers have also been successfully designed and used. The *model predictive control* is typically employed for this purpose, where the controller constantly evolves (or refines) control laws using an integrated simulation model of the controlled system. A comprehensive description of this approach is given in (Onnen et al., 1997).

3.2. Flight Control For The Uav Recovery Task

Aircraft control at recovery stage of flight can be conventionally separated into two closely related, but distinctive tasks: guidance and flight control. Guidance is the high-level ('outer loop') control intended to accomplish a defined mission. This may be path following, target tracking or various navigation tasks. Flight control is aimed at providing the most suitable conditions for guidance by maintaining a range of flight parameters at their optimal levels and delivering the best possible handling characteristics.

The landing of an aircraft is a well established procedure which involves following a predefined flight path. More often than not, this is a rectilinear trajectory on which the aircraft can be stabilised, and the control interventions are needed only to compensate disturbances and other sources of errors.

The position errors with respect to the ideal glidepath can be measured relatively easily. Shipboard landing on air carriers has tight error tolerances and absence of flare manoeuvres before touchdown. The periodic ship motion does have an effect on touchdown; however, it does not affect significantly the glidepath, which is projected assuming the average deck position.

Ship oscillations in high sea cause periodic displacement of the recovery window makes it impossible to project an optimal flight path when the final approach starts. Therefore, it turns out that the UAV recovery problem resembles that of homing guidance rather than typical landing. While stabilisation on a known steady flight path can be done relatively easy with a PID controller, homing guidance to a moving target often requires a more sophisticated control.

3.3. Uav Controller Structure

The objective is to synthesise such guidance strategy that enables reliable UAV recovery, and to produce a controller that implements the target tracking guidance strategy.

The evolutionary design (ED) method applied for this task allows to evolve automatically both the structure and the parameters of the control laws, thus potentially enabling to generate a 'full' controller, which links available measurements directly with the aircraft control inputs (throttle, ailerons, rudder and elevator) and implements both the guidance strategy and flight control (Fig. 6):

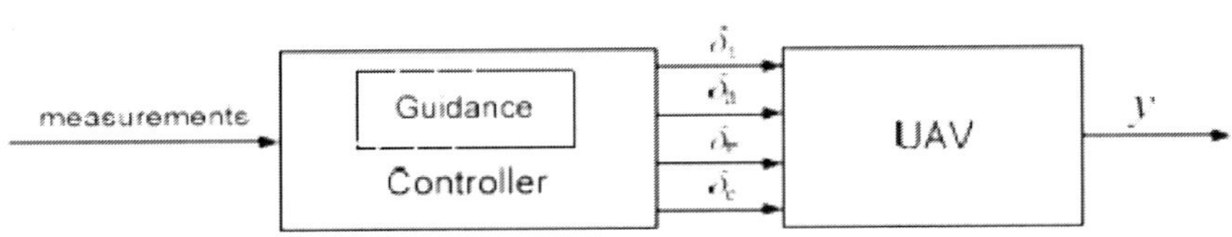

Figure 6. Full controller with embedded guidance strategy

However, this approach, even though appealing at first and requiring minimum initial knowledge, proves to be impractical as the computational demands of the evolutionary algorithms (EAs) soar exponentially with the dimensionality of the problem. It is therefore desirable to reduce complexity of the problem by reducing the number of inputs/outputs and limiting, if appropriate, possible structures of the controllers.

Also it is highly desirable to decompose the complex control task into several simpler problems and to solve them separately. A natural way of such decomposition is separating the trajectory control (guidance) and flight control. The guidance controller issues commands u_g to the flight controller, which executes these commands by manipulating the control surfaces of the UAV (Fig. 7). These two controllers can be synthesised separately using appropriate fitness evaluation for each case.

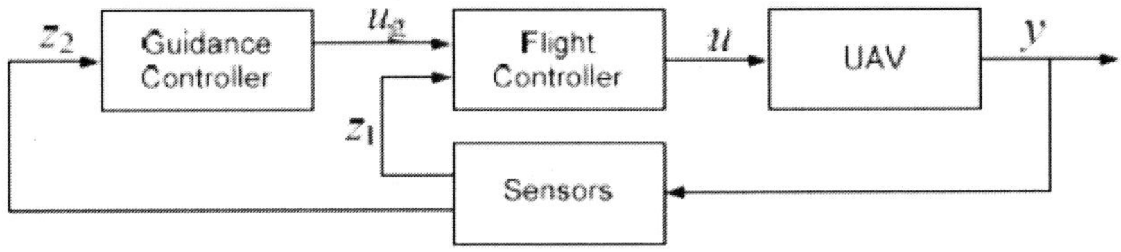

Figure 7. UAV recovery control diagram

The internal structure of the controller is defined by the automatic evolutionary design based on predefined set of inputs and outputs. As an additional requirement is that the outputs u_g should represent a group of measurable flight parameters such as body accelerations, velocities and Euler angles, which the flight controller can easily track.

The structure of the output part of the guidance controller is as shown in Fig. 8 which allows to evolve the horizontal and vertical guidance laws separately, which may be desirable due to different dynamics of the UAV's longitudinal and lateral motion and also due to computational limitations.

Input measurements to the guidance controller should be those relevant to trajectory like positioning information, pitch and yaw angles and airspeed.

The derived quantities from raw measurements are the vertical and lateral velocity components with respect to the approach ground reference frame.

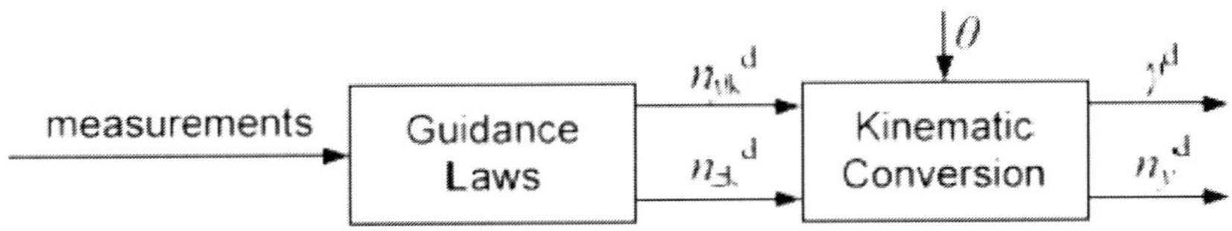

Figure 8. Guidance controller

The system is based on radio distance metering and provides ten independent raw measurements (Fig. 9): three distances d_1, d_2 and d_3 from the UAV to the radio transmitters located at both ends of the recovery boom which supports the arresting wire and at the base of recovery mast; three rates of change of these distances; distance differences $(d_1 - d_2)$ and $(d_3 - d_2)$; and rates of change of the differences.

The guidance laws evolution process is potentially capable to produce the laws directly from raw measurements, automatically finding necessary relationships between the provided data and the required output.

Flight controller receives two inputs from the guidance controller: bank angle demand γ^d and normal body load factor demand ndy. It should track these inputs as precisely as possible by manipulating four aircraft controls: throttle, ailerons, rudder and elevator.

The available measurements from the onboard sensors are body angular rates ω_x, ω_y, ω_z from rate gyros, Euler angles γ, ψ, θ, body accelerations n_x, n_y, n_z from the respective accelerometers, airspeed V_a, aerial angles α and β, actual deflection of the control surfaces δ_a, δ_r, δ_e, and engine rotation speed N_{rpm}.

For simplicity of design, the controller is subdivided into independent longitudinal and lateral components. In longitudinal branch, elevator tracks the input signal ndy, while throttle is responsible for maintaining a required airspeed. In lateral control, naturally, ailerons track γ^d, while rudder minimises sideforce by keeping n_z near zero.

Figure 9. Positioning scheme

4. EVOLUTIONARY DESIGN

The Evolutionary Design (ED) presented in this section, generally, takes no assumptions regarding the system and thus can be used for wide variety of problems, including nonlinear systems with unknown structure. The core of evolutionary design is a specially tailored evolutionary algorithm (EA) which evolves both the structure and parameters of the control laws.

Parallel evolution can be implemented in a variety of ways. One of the few successfully employed variants is the *block structure controller evolution* (Koza et al., 2000).

In this work the ED algorithm enables to evolve suitable control laws within a reasonable time by utilising gradual evolution with the principle of strong casualty. This means that structure alterations are performed so that the information gained so far in the structure of the control law is preserved. Addition of a new block, though being random, does not cause disruption to the structure. Instead, it adds a new dimension and new potential which may evolve later during numerical optimisation.

The addition of new points or blocks is carried out as a separate dedicated operation (unlike sporadic structure alterations in the sub-tree crossover), and is termed *structure mutation*. Furthermore, in this work structure mutation is performed in a way known as *neutral structure mutation*. That's when the new block should be placed initially with zero coefficient. The usefulness of neutral mutations has been demonstrated for the evolution of digital circuits (Van Laarhoven & Aarts, 1987) and aerodynamic shapes (Olhofer et al., 2001).

As a result, the ED algorithm basically represents a numerical EA with the inclusion of structure mutations mechanism.

Control laws are represented as a combination of static functions and input signals, which are organised as a dynamic structure of state equations and output equations in form of continuous representation.

The controller being evolved has m inputs, r outputs and n states. So, the controller comprises of n state equations and r output equations:

$$\begin{matrix} \dot{x}_1 = g_1(x,u) \\ \dot{x}_2 = g_2(x,u) \\ \vdots \\ \dot{x}_n = g_n(x,u) \end{matrix} \quad \text{and} \quad \begin{matrix} y_1 = f_1(x,u) \\ y_2 = f_2(x,u) \\ \vdots \\ y_n = f_n(x,u) \end{matrix} \tag{9}$$

where u is size m vector of input signals, $x = [x_1, x_2, \ldots x_n]$ is size n vector of state variables, $y_{1 \ldots r}$ are controller outputs. Initial value of all state variables is zero. All n+r equations are built on the same principle and are evolved simultaneously. For structure mutations, a random equation is selected from this pool and mutated.

Input signals delivered to each particular controller are directly measured signals as well as the quantities derived from them. Within each group, inputs are organised in the subgroups of 'compatible' parameters. Compatible parameters are those which have close relationship with each other, have the same dimensions and similarly scaled. The examples of compatible parameters are the pairs (n_x, n_{xg}), (ω_y, ψ), (V_a, V_{CL}).

Therefore, every controller input may be represented by a unique code consisting of three indices: the number of group a, the number of subgroup b and the number of item in the subgroup c. The code is designated as u(a,b,c).

Each of the control equations (9) is encoded as described above. To this end, only one single output equation of the form y = f(u) will be considered in this section.

The encoding should allow a simple way to insert a new parameter in any place of the equation without disrupting its validity and in a way that this insertion initially does not affect the result, thus allowing neutral structure mutations. Conceptually, the equation is a sum of input signals, in which:

- every input is multiplied by a numeric coefficient or another similarly constructed expression;
- the product of the input and its coefficient (whether numeric or expression) is raised to the power assigned to the input;
- a free (absolute) term is present.

The simplest possible expression is a constant:

$$y = k_0 \tag{10}$$

A linear combination of inputs plus a free term is also a valid expression:

$$y = k_2u_2 + k_1u_1 + k_0 \tag{11}$$

Any numeric constant can be replaced with another expression. An example of a full featured equation is

$$y = ((k_4u_4 + k_3)u_3)^{-0.5} + k_2u_2 + (k_1u_1)^2 + k_0 \tag{12}$$

This algorithm can be illustrated by encoding the example (12). The respective internal representation of this expression is:

- Equation: $y = ((k_4u_4 + k_3)u_3)^{-0.5} + k_2u_2 + (k_1u_1)^2 + k_0$
- Expression: { u(3,-0.5) u(4) 1 2 u(2) 3 u(1,2) 4 5 }
- Object parameters: [k_4 k_3 k_2 k_1 k_0]
- Strategy parameters: [s_4 s_3 s_2 s_1 s_0]

This syntax somewhat resembles Polish notation with implicit '+' and '*' operators before each variable. The representation ensures presence of a free term in any sub-expression, such as k_3 and k_0 in the example above.

The algorithm of structure mutation is presented below.

1 Select an input (or a state variable) at random: u(a,b,c).

2 Obtain the initial values of the numeric coefficient and the strategy parameter

3 Append the object parameters vector with the initial coefficient, and the strategy parameters vector with the initial step size.

4 Form a sub-expression consisting of the selected variable code and the obtained index: { u(a,b,c) n }.

5 With 40% probability, set the insertion point (locus) to 1; otherwise, select a numeric value in the expression at random with equal probability among all numeric values present and set the locus to the index of this value.

6 Insert the sub-expression into the original expression at the chosen locus (*before* the item pointed).

This procedure may produce redundant expressions when the selected variable already exists at the same level. Thus an algebraic simplification procedure is implemented.

Fitness evaluation of a controller can be divided into two main stages. First is the preparation of the sample task and simulation of the model with the controller in the loop. The second stage is analysis of the results obtained from the simulation and evaluation of the fitness as such.

Other parameters of flight taken into account is control usage (or control effort): it is desirable to keep control usage at minimum.

The total fitness value is calculated as the weighted sum of all estimates:

$$F = W_c C_c + W_f C_f + W_e C_e + \ldots \tag{13}$$

The exact value of the weighting coefficients W, as well as the number of estimates taken into account, is individual for each controller. As a rule, the weighting coefficients are chosen empirically.

Several steps of design algorithm need further clarification.

Initial population is initialised with the control laws of the form $y = const$ with randomly chosen constants.

The selection is performed deterministically. The populations used in this study were of moderate size, usually 24 to 49 members. Selection pressure is determined by the number of the offspring n of each individual. The smaller n, the lower the selection pressure. For nearly all runs in this work n = 2, which means that half of the population is selected. This is a rather mild level of selection pressure.

The parameters determining *structure mutation* occurrence, k_s and P_s, both change during the evolution. The probability of structure mutation P_s is normally high in the beginning (0.7 to 1.0) and then decrease exponentially to moderate levels (0.4 to 0.6) with the exponent 0.97 to the generation number. In the beginning of evolution, k_s is reduced by half until the generation 20 and 100 respectively.

Reproduction is performed simultaneously with mutation, as it is typically done in ES, with the exception that this operation is performed separately for each selected member.

5. CONTROLLER SYNTHESIS AND TESTING

The UAV control system is synthesised in several steps. First, the flight controller is produced. This requires several stages, since the flight controller is designed separately for longitudinal and lateral channels. When the flight controller is obtained, the guidance control laws are evolved.

Application of the ED algorithm to control laws evolution is fairly straightforward: 1) preparation of the sample task for the controller, 2) execution of the simulation model for the given sample task and 3) analysis of the obtained performance and evaluation of the fitness value. For a greater detail of control system design reader is referred to (Bourmistrova & Khantsis, 2009). When both the model and fitness evaluation are prepared, the final evolution may be started. Typically, the algorithm is run for 100–200 generations (depending on complexity of the controller being evolved). The convergence and the resulting design is then analysed and the evolution, if necessary, is continued.

Step 1. PID autothrottle - Initially a simple PID variant of autothrottle is evolved - to ensure a more or less accurate airspeed hold. At the next stage, its evolution is continued in a full form together with the elevator control law. The PID structure of the controller may be ensured by appropriate initialisation of the initial population and by disabling structure mutations. Therefore, the algorithm works as a numerical optimisation procedure. The structure of the autothrottle control law is following:

$$\begin{aligned} \dot{x}_1 &= k_1 x_1 + k_2 \Delta V_a \\ \delta_t &= k_3 x_1 + k_4 \Delta V_a + k_5 \dot{V}_a + k_6 \end{aligned} \tag{14}$$

where $\Delta V_a = V^d - Va$ is the airspeed error signal, δt is the throttle position command and $k_{1...6}$ are the coefficients to be optimised.

Step 2. *Longitudinal control* - With a simple autothrottle available, elevator control can be developed to provide tracking of the normal body load factor demand ndy. Altogether, the reference input signal is

$$n_y^d(t) = 0.03\left(H^d - H(t)\right) + \cos\theta(t) + 0.2c(t) \quad (15)$$

Elevator control law is initialised as follows:

$$\begin{aligned} \dot{x}_2 &= 0 \\ \delta_e &= k_1 x_2 + k_2 \Delta n_y + k_3 \end{aligned} \quad (16)$$

where $\Delta n_y = n_y^d - n_y$ is the load factor error and the coefficients $k_{1...3}$ are sampled at random for the initial population.

Step 3. *Lateral control* - Lateral control consists of two channels: ailerons control and rudder control. As a rule, for an aerodynamically stable aircraft such as the Ariel UAV, lateral control is fairly simple and is not as vital for flight as longitudinal control. For this reason, both control laws, for ailerons and rudder, are evolved simultaneously in one step.

Both ailerons and rudder control laws are initialised in a similar manner to (16):

$$\begin{aligned} \dot{x}_3 &= 0 \\ \dot{x}_4 &= 0 \\ \delta_a &= k_{11} x_3 + k_{12} \Delta\gamma + k_{13} \\ \delta_r &= k_{21} x_4 + k_{22} n_z + k_{23} \end{aligned} \quad (17)$$

Step 4. *Guidance* - At this point, flight controller synthesis is completed and guidance laws can be evolved. Guidance controller comprises two control laws, for the vertical and horizontal load factor demands, n_{yk}^{d} and n_{zk}^{d} respectively. This is equivalent to acceleration demands

$$a_{yk}^{d} = g n_{yk}^{d} \text{ and } a_{zk}^{d} = g n_{zk}^{d} \tag{18}$$

In the guidance task, the random factors include initial state of the UAV; Sea State, atmospheric parameters, and initial state of the ship.

Fitness is calculated as follows:

$$F_g = 40\Delta h_1^2 + 20\Delta z_1^2 + 50|\psi_1| + 25|\gamma_1| + 500C_c\left(n_{yk}^{d}\right) + 500C_c\left(n_{zk}^{d}\right) + 100C_f\left(n_{zk}^{d}\right)$$

(19)

where Δh1 and Δz1 are vertical and horizontal miss distances, and ψ1 and γ1 are final yaw and bank angles. Greater weight for vertical miss than that for horizontal miss is used because vertical miss allowance is smaller (approximately 3–4 m vs. 5 m) and also because vertical miss may result in a crash into the boom if the approach is too low.

Algorithm initialisation is the same as for longitudinal flight control laws evolution, except that elitism is not used and the population size is 48 members. The initial population is sampled with the control laws of the form

$$\begin{aligned} \dot{x}_6 &= 0 \\ \dot{x}_7 &= 0 \\ a_{yk}^{d} &= k_{11} x_6 + k_{12} \\ a_{zk}^{d} &= k_{21} x_7 + k_{22} \end{aligned} \tag{20}$$

where all coefficients k are chosen at random. For convenience, the control laws are expressed in terms of accelerations, which are then converted to load factors according to (18). Since these control laws effectively represent 'empty' laws y = *const* (plus a low-pass output filter with randomly chosen bandwidth), structure mutation is applied to each member of the initial population.

From the final populations, the best solution is identified by calculating fitness of each member using N = 25 simulation runs, taking into account also success rate. As an eample the best performing controller is the following (unused state variables are removed) with 100% success rate in the 25 test runs with average fitness 69.52:

$$\begin{aligned} \dot{x}_6 &= -3.548\omega_v \\ a_{yk}^d &= 0.916x_6 - 54.3\omega_v - 0.1261 \\ a_{zk}^d &= -107.68\omega_h + 0.0534V_{zT} + 0.4756 \end{aligned} \tag{21}$$

The other attempted approach is the pure proportional navigation (PPN) law for recovery which was compared with the obtained solutions (Duflos et al., 1999, Siouris & Leros, 1988). The best PN controller have been selected using the fitness values of each member of the final population averaged over 100 simulation runs. It is the following:

$$\begin{aligned} a_{yk}^d &= -3.27V_{CLa}\omega_V \\ a_{zk}^d &= -3.18V_{CLa}\omega_h \end{aligned} \tag{22}$$

This controller received the fitness 88.85 and showed success rate 95%.

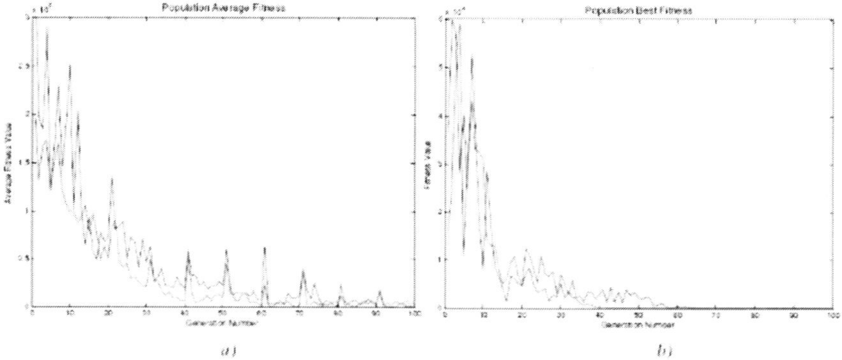

Figure 10. Population average fitness (a) and best fitness (b) for two independent evolutions of guidance controller

5.1. Controller Testing

In this work, two main types of simulation tests are conducted – robustness and performance. Results are presented for robustness test, which is aimed at ensuring the controller has good robustness to modelling uncertainties and to test whether the controller is sensitive to specific perturbations.

Perturbations were introduced in two ways - in form of variation of physical quantities and by introducing additional dynamic elements and by changing internal variables such as aerodynamic coefficients. For a realistic test, all perturbations should be applied simultaneously to identify the worst case scenario. However, single perturbation tests (the *sensitivity analysis*) allow to analyse the degree of influence of each parameter and help to plan the robustness test more systematically.

In this type of test, a single model variable is perturbed by a set amount and the effect upon the performance of the controller is determined. Performance evaluation can be measured in a manner similar to fitness evaluation of the guidance controller (equation (19)).

The additional parameters taken into account in performance measurement are impact speed V_{imp} and minimum altitude H_{min} attained during the approach.

Altogether, the performance cost (PC) is calculated as follows:

$$PC = 40\Delta h_1^2 + 20\Delta z_1^2 + 50|\psi_1| + 25|\gamma_1| + 50 f_V + 20 f_H + 10C_c(\delta_a) + 10C_c(\delta_r) + C_c(\delta_e) + \ldots$$
$$+ 200C_f(\delta_a) + 200C_f(\delta_r) + 200C_f(\delta_e)$$

(23)

where

$$f_V = \begin{cases} V_{imp} - 30, & V_{imp} > 30 \\ 0, & V_{imp} \le 30 \end{cases}, \quad f_H = \begin{cases} 10 - H_{min}, & H_{min} > 10 \\ 0, & H_{min} \le 10 \end{cases}. \quad (24)$$

Impact speed V_{imp} and minimum altitude H_{min} are measured in m/s and metres respectively. Other designations are as in (19). Unlike fitness evaluation in the flight controller evolution, the commanded control deflections δa, δr and δe are saturated as required for control actuators.

The environment delivers a great deal of uncertainty. The range of disturbances and the initial ship phase are chosen to provide a moderately conservative estimation of controller performance. Results for different dynamic scenarios are presented in (Bourmistrova & Khantsis, 2009).

The parameters corresponding to aircraft geometry and configuration are tested in a similar manner. The range is increased to the scale factors between 0 and 10 (0 to 1000%) with step 0.05.

The allowable perturbations (as factors to the original values) are summarised in Table 1, where * denotes the extreme value tested.

Table 1. Allowable perturbations of UAV inertial properties and geometry

Parameter	Lower limit factor		Upper limit factor	
	Controller 1	Controller 2	Controller 1	Controller 2
Empty mass (m_{empt})	0.50*	0.50*	1.54	1.52
Rolling moment of inertia (I_x)	0.20	0.20	10*	2.43
Yawing moment of inertia (I_y)	0*	0*	10*	10*
Pitching moment of inertia (I_z)	0*	0*	2.17	1.91
xy cross product of inertia (I_{xy})	0*	0*	10*	10*
Wing area (S)	0.74	0.87	1.55	1.60

Example in Fig. 11 demonstrates results for varying empty mass. Dashed cyan and green lines on the trajectory graphs represent the 'unrolled' along the flight path traces of the tips of the recovery boom (lateral position on the top view and vertical position on the side view). The bar on the right hand side illustrates the size of recovery window.

For some parameters, no noticeable drop in performance is experienced within the testing limits. These limits indicate quite large perturbations that can reasonably be expected. The main perturbations that cause a performance degradation are those that affect the physically attainable trajectory and not the operation of the controller, e.g. increasing weight and decreasing wing area. However, as follows from the above analysis, careful adjustment of elevator efficiency and/or horizontal stabiliser incidence may help to increase the tolerance to increased weight. There is still sufficient margin in angles of attack and engine power.

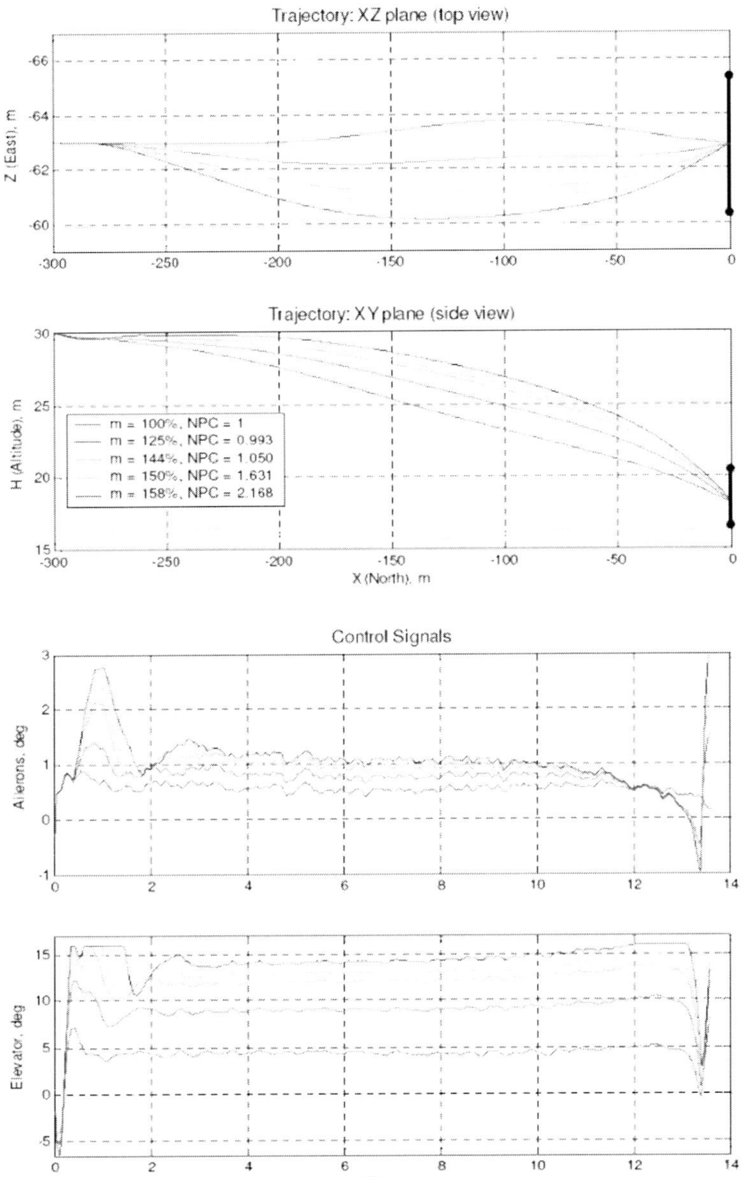

Figure 11. Flight path and control signals with varying empty mass

The other parameters evaluated in this research were the perturbations of the power unit parameters, the aircraft aerodynamics parameters and sensor noise (Khantsis, 2006).

Overall, these tests have not exposed any significant robustness problems within the controllers. Large variations in single aircraft parameters caused very few control problems. However, such variations cannot realistically judge the performance of the controllers under simultaneous perturbations of multiple parameters.

6. CONCLUSIONS

In this chapter, an application of the Evolutionary Design (ED) is demonstrated. The aim of the design was to develop a controller which provides recovery of a fixed-wing UAV onto a ship under the full range of disturbances and uncertainties that are present in the real world environment.

Evolutionary computation is an attractive and quickly developing technique. Methodological shortcomings of the early approaches and lack of intercommunication between different EA schools contributed to the fact that evolutionary computation remained relatively unknown to the engineering and scientific audience for almost three decades. Despite computational demands, evolutionary methods offer many advantages over conventional optimisation and problem solving techniques. The most significant one is flexibility and adaptability of EAs to the task at hand.

In this study, a combination of the EA methods is used to evolve a capable UAV recovery controller. An adapted GP approach is used to represent the control laws. However, these laws are modified more judiciously (yet stochastically) than commonly accepted in GP and evolved in a manner similar to ES approach.

One of the greatest advantages of developed methodology is that minimum or no a priori knowledge about the control methods is used, with the synthesis starting from the most basic proportional control or even from 'null' control laws. During the evolution, more complex and capable laws emerge automatically. As the resulting control laws demonstrate, evolution does not tend to produce parsimonious solutions.

The method demonstrating remarkable robustness in terms of convergence indicating that a near optimal solution can be found. In very limited cases, however, it may take too long time for the evolution to discover the core of a potentially optimal solution, and the process does not converge. More often than not, this hints at a poor choice of the algorithm parameters.

The simulation testing covers the entire operational envelope and highlights several conditions under which recovery is risky. All environmental factors—sea wave, wind speed and turbulence—have been found to have a significant effect upon the probability of success. Combinations of several factors may result in very unfavourable conditions, even if each factor alone may not lead to a failure. For example, winds up to 12 m/s do not affect the recovery in a calm sea, and a severe ship motion corresponding to Sea State 5 also does not represent a serious threat in low winds. At the same time, strong winds in a high Sea State may be hazardous for the aircraft.

On the whole, Evolutionary Design is a useful and powerful tool for complex nonlinear control design. Unlike most other design methodologies, it tries to solve the problem at hand automatically, not merely to optimise a given structure. Although ED does not exclude necessity of a thorough testing, it can provide a near optimal solution if the whole range of conditions is taken into account in the fitness evaluation. In principle, no specific knowledge about the system is required, and the controllers can be considered as 'black boxes' whose internals are unimportant. Successful design of the controller for such a challenging task as shipboard recovery demonstrates great potential abilities of this novel technique.

REFERENCES

1. H. G. Beyer, K. Deb, 2001 On self-adaptive features in real-parameter evolutionary algorithms, Proceedings on IEEE Transactions on Evolutionary Computation, 250 270 5 3), 1089 Germany, June 2001

2. A. Bourmistrova, 2001 Knowledge based control system design for autonomous flight vehicle, PhD thesis, RMIT University: Melbourne, September 2001

3. A. Bourmistrova, S. Khantsis, 2009 Flight control system design optimisation via Genetic Programming, in Recent advances in signal processing, IN-TECH, 978-9-53761-941-1

4. A. Chipperfield, P. Fleming, 1996 Systems integration using evolutionary algorithms, Proceedings of UKACC International Conference on Control'96, 705 710 1 October 1996

5. D. A. Coley, 1998 An Introduction to Genetic Algorithms for Scientists and Engineers. University of Exeter, 9-81023-602-6 by World Scientific, 1999

6. K. A. De Jong, W. M. Spears, On the state of evolutionary computation, ed. S. Forrest, 5 5th International Conference on Genetic Algorithms proceedings. San Mateo, CA, Morgan Kaufmann: 618 1993.

7. N. L. Cramer, 1985 A representation for the adaptive generation of simple sequential programs, Proceedings on International Conference on Genetic Algorithms and their Applications, 183

187 , ed. J. J. Grefenstette, Carnegie-Mellon University, Pittsburgh May 1985

8. K. Deb, D. Joshi, A. Anand, 2001 Real-coded evolutionary algorithms with parentcentric recombination, KanGAL Report 2001003, Indian Institute of Technology.
9. E. Duflos, P. Penel, P. Vanheeghe, 1999 3D guidance law modelling, IEEE Transactions on Aerospace and Electronic Systems, 35 1 page numbers (72-83), (January 1999), 0018-9251
10. L. J. Eshelman, J. D. Schaffer, 1993 Real-coded genetic algorithms and interval schemata, in Foundations of Genetic Algorithms 2, page numbers (187-202), ed. L. D. Whitley, Morgan Kaufmann: San Mateo, CA, 1993
11. D. B. Fogel, 1962 Autonomous automata, Industrial Research, 4 page numbers (14-19), 1962
12. D. B. Fogel, A. J. Owens, M. J. Walsh, 1966 Artificial intelligence through simulated evolution. New York: Wiley, 1966
13. D. E. Goldberg, 1989 Genetic Algorithms in search, optimisation and machine learning. Addison-Wesley, 1989
14. D. E. Goldberg, K. Deb, 1991 A comparative analysis of selection schemes used in genetic algorithms, Foundations of Genetic Algorithms, page numbers (69-93), ed. G. J. E. Rawlins. Morgan Kaufmann, San Mateo, CA, 101558601708
15. J. J. Grefenstette, 1999 Evolvability in dynamic fitness landscapes: A genetic algorithm approach, Proceedings on Congress of Evolutionary Computation, IEEE Press, 2031 2038 3 Washington, 0-78035-536-9
16. F. Herrera, M. Lozano, J. L. Verdegay, 1998 Tackling real-coded genetic algorithms: Operators and tools for behavioural analysis,

Artificial Intelligence Review, page numbers (265-319), 12 4 Springer, 0269-2821

17. J. H. Holland, 1975 Adaptation in Natural and Artificial Systems. University of Michigan Press

18. R. B. Hollstien, 1971 Artificial genetic adaptation in computer control systems, Doctoral dissertation, University of Michigan

19. N. Kaise, Y. Fujimoto, 1999 Applying the evolutionary neural networks with genetic algorithms to control a rolling inverted pendulum, Lecture Notes in Computer Science : Simulated Evolution and Learning, 1585 1999, (January 1999), page numbers (223-230), Springer Berlin / Heidelberg, 978-3-54065-907-5

20. S. Khantsis, 2006 Control System Design Using Evolutionary Algorithms for Autonomous Shipboard Recovery of Unmanned Aerial Vehicles, PhD thesis, RMIT University: Melbourne, 1992

21. J. R. Koza, 1992 Genetic Programming: On the programming of computers by means of natural selection, Cambridge, Massachusetts: MIT Press, December 1992 (6th ed. 1998), 100262111705

22. J. R. Koza, 1994 Genetic Programming II: Automatic discovery of reusable programs. Cambridge, Massachusetts: MIT Press, May 1994, 100262111896

23. J. R. Koza, I. I. I. F. H. Bennett, D. Andre, M. A. Keane, 1996 Four problems for which a computer program evolved by genetic programming is competitive with human performance, Proceedings on IEEE International Conference on Evolutionary Computation, Nagoya, Japan, May 1996, 0-78032-902-3

24. J. R. Koza, I. I. I. F. H. Bennett, D. Andre, M. A. Keane, 1999 Genetic Programming III: Darwinian Invention and Problem Solving. Kluwer Academic Publishers, 1-55860-543-6

25. J. R. Koza, M. A. Keane, J. Yu, I. I. I. F. H. Bennett, W. Mydlowec, 2000 Automatic creation of human-competitive programs and controllers by means of genetic programming, Genetic Programming and Evolvable Machines, 1 1 page numbers (121-164), (January 2000), 1389-2576

26. J. R. Koza, M. A. Keane, M. J. Streeter, W. Mydlowec, J. Yu, G. Lanza, 2003 Genetic Programming IV: Routine Human-Competitive Machine Intelligence. Kluwer Academic Publishers

27. W. B. Langdon, 1999 The evolution of size and shape, in Advances in genetic programming 3, ed. L. Spector, W. B. Langdon and et al, MIT Press, page numbers (163-190), July 1999, 100262194236

28. W. B. Langdon, R. Poli, 2002 Foundations of genetic programming. Berlin: Springer-Verlag, 978-3-54042-451-2

29. S. W. Mahfoud, G. Mani, 1996 Financial forecasting using genetic algorithms. Applied Artificial Intelligence, page numbers (543-565), 10 November 1996, 1087-6545

30. R. E. Mc Farland, K. Duisenberg, 1995 Simulation of rotor blade element turbulence, Technical Memorandum 108862, Acc. No. NASA TM-108862, NASA, January 1995

31. D. Mc Lean, H. Matsuda, 1998 Helicopter station-keeping: comparing LQR, fuzzy-logic and neural-net controllers, Engineering Applications of Artificial Intelligence, page numbers (411- 41811), 11 November 1998, 0952-1976

32. P. Nordin, W. Banzhaf, 1995 Complexity compression and evolution, ed. L. J. Eshelman, Proceedings on 6th International

Conference on Genetic Algorithms, 310 317 , Pittsburgh, PA, Morgan Kaufmann, July 1995

33. M. Olhofer, Y. Jin, B. Sendhoff, 2001 Adaptive encoding for aerodynamic shape optimization using Evolution Strategies, Proceedings of Congress on Evolutionary Computation, 576 583 1 Seoul, South Korea, 2001, 1-59593-010-8

34. C. Onnen, R. Babuska, U. Kaymak, J. M. Sousa, H. B. Verbruggen, R. Isermann, 1997 Genetic algorithms for optimization in predictive control, Control Engineering Practice, 5 10 page numbers (1363-1372), (October 1997), 0967-0661

35. I. Ono, S. Kobayashi, 1997 A real-coded genetic algorithm for function optimisation using unimodal normal distribution crossover, ed. T. B$ck, Proceedings on 7th International Conference on Genetic Algorithms, 246 253 , San Mateo, CA, Morgan Kauffman, 1997

36. T. Perkins, 1994 Stack-based genetic programming, Proceedings on IEEE World Congress on Computational Intelligence, 148 153 1 Orlando, Florida, IEEE Press, 27-29 June 1994

37. H. P. Schwefel, 1981 Numerical optimisation of computer models. New York: Wiley

38. B. Sendhoff, M. Kreuz, 1999 Variable encoding of modular neural networks for time series prediction, ed. V. W. Porto, Proceedings on Congress on Evolutionary Computation, 259 266 , IEEE Press, Piscataway, NJ, July 1999

39. G. M. Siouris, P. Leros, 1988 Minimum-time intercept guidance for tactical missiles, Control Theory and Advanced Technology, 4 2 page numbers (251-263), (February, 1988), 0911-0704

40. S. F. Smith, 1980 A learning system based on genetic adaptive algorithms, PhD dissertation, University of Pittsburgh

41. W. M. Spears, V. Anand, 1991 A study of crossover operators in genetic programming, ed. Z. W. Ras, M. Zemankova, Proceedings on International Symposium on Methodologies for Intelligent Systems, 409 418 , Berlin: Springer-Verlag, 542., 1991, 3-54054-563-8

42. R. K. Ursem, 2003 Models for evolutionary algorithms and their applications in system identification and control optimization, PhD dissertation, Department of Computer Science, University of Aarhus, Denmark, 2003

43. P. J. M. van Laarhoven, E. H. L. Aarts, 1987 Simulated annealing: theory and applications. Dordrecht, The Netherlands: D. Reidel, 109027725136

44. Q. Wang, A. M. S. Zalzala, 1996 Genetic control of near time-optimal motion for an industrial robot arm, Proceedings of IEEE International Conference on Robotics and Automation, 2592 2597 3 0-78032-988-0 MN, April 1996

CHAPTER 4

CLOSED-LOOP DYNAMIC CONTROL ALLOCATION FOR AIRCRAFT WITH MULTIPLE ACTUATORS

Wendong Gai , Honglun Wang,

Science and Technology on Aircraft Control Laboratory, Beihang University, Beijing 100191, China

ABSTRACT

A closed-loop control allocation method is proposed for a class of aircraft with multiple actuators. Nonlinear dynamic inversion is used to design the baseline attitude controller and derive the desired moment increment. And a feedback loop for the moment increment produced by the deflections of actuators is added to the angular rate loop, then the error between the desired and actual moment increment is the input of the dynamic control allocation. Subsequently, the stability of the closed-loop dynamic control allocation system is analyzed in detail. Especially, the closed-loop system stability is also analyzed in the presence of two types of actuator failures: loss of effectiveness and lock-in-place actuator failures, where a fault detection subsystem to identify the actuator failures is absent. Finally, the proposed method is applied to a canard rotor/wing (CRW) aircraft model in fixed-wing mode, which has multiple actuators for flight control. The nonlinear simulation demonstrates that this method can guarantee the stability and tracking performance whether the actuators are healthy or fail.

KEYWORDS

Canard rotor/wing aircraft; Closed-loop control allocation; Dynamic inversion; Flight control systems; Redundant actuators

1. INTRODUCTION

Modern aircraft, automotive vehicles and marine vessels are usually equipped with more control actuators than controlled variables to achieve multiple control objectives and high performance. And the nonlinear control design methods, like dynamic inversion[1] and [2]and backstepping,[3] and [4] result in control laws specifying the forces and moments, rather than the control surface deflections. Determining how to distribute the control signals to the available actuators is known as the control allocation problem. Meanwhile, the actuator selection is separated from the regulation task to simplify the control design using the control allocation. In particular, an effective re-allocation among the remaining healthy actuators can maintain the acceptable performance in the case of actuator failures.

Control allocation methods have been studied extensively. The survey paper that compares the strengths and limitations of control allocation methods is presented in Ref.[5]. Regardless of methods, such as pseudoinverse,[6] direct allocation,[7] and [8] daisy chain allocation,[9] linear programming[10] and nonlinear programming,[11] the resulting methods are static in the sense that the control input depends only on the current virtual control command.

Different from these methods, the dynamic control allocation method[12] and [13] is proposed to make use of the redundancy to get different actuators operate in the different parts of frequency domain. The control allocation mapping of this method is a linear filter when there are no actuators in saturation state. And the control designer can decide the frequency characteristic of this filter by weighting matrices selected.

Both the static and the dynamic methods mentioned above are the open-loop control allocation. These control allocation approaches can be implemented to satisfy the optimization criterion when there is no actuator failure. However, the system stability or tracking performance becomes a problem if some unknown actuator failures occur.

Recently, a new nonlinear flight control design method is presented in Ref.[14] for aircraft with redundant actuators, combining the bases sequence control allocation with the moment compensation to implement the desired moment commands. However, this method fails to consider the closed-loop stability of control allocation system whether the actuators are healthy or fail. In this paper, a new systematic method for closed-loop dynamic control allocation is proposed, and the design process and stability analysis are introduced in detail.

2. CLOSED-LOOP DYNAMIC CONTROL ALLOCATION

As mentioned in the first section, the control allocation plays an important role in nonlinear flight control for aircraft with multiple actuators, particularly, when the actuators have dynamics, limits and failures. A conceptual block diagram of the attitude control loop with closed-loop dynamic control allocation is shown in Fig. 1. The moments $\boldsymbol{M}$acting on the aircraft are given as

equation(2)

$$\boldsymbol{M} = \boldsymbol{M}_0 + \Delta \boldsymbol{M}$$

where $\boldsymbol{M}_0$ is determined by the aircraft configuration and flight states, and $\Delta\boldsymbol{M}$ the moment increments produced by the control surface deflections.

The control allocation problem is solved by the moment allocation among different actuators in this research. In Fig. 1, the input of system is the commanded attitude $\boldsymbol{r}_g$. And the desired moment increment $\Delta\boldsymbol{M}_d$ is derived by the nonlinear dynamic inversion attitude controller in outer-loop, while the actual moment increment $\Delta\boldsymbol{M}$ can be measured by the angular acceleration sensors. Then, the error $\boldsymbol{v}$ between the desired moment increments $\Delta\boldsymbol{M}_d$ and the actual moment increments $\Delta\boldsymbol{M}$ is as the input of dynamic control allocation. Thus, the output of control allocation $\boldsymbol{uv}$ has the form of increment and one integration step delay is required. At last, the input of actuator is $\boldsymbol{u}(k) = \boldsymbol{u}(k-1) + \boldsymbol{uv}(k)$, and k denotes the current sampling period. And the output of actuator $\boldsymbol{\delta}$ is deflections of actuators, which produces the actual moment increments $\Delta\boldsymbol{M}$ and aerodynamic force $\boldsymbol{F}$.

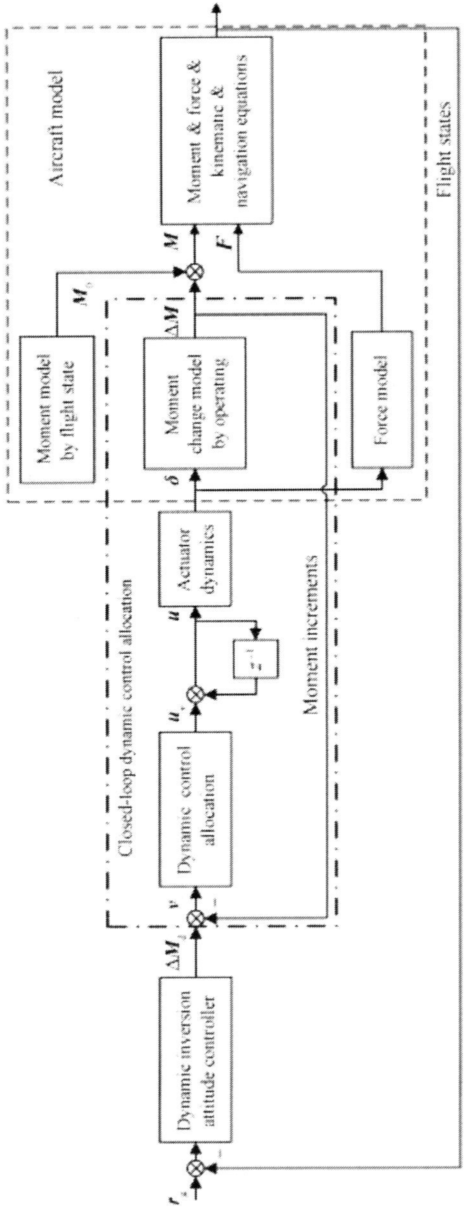

Figure. 1. Closed-loop dynamic control allocation configuration.

2.1. Aircraft model

The aircraft model is described as[15]

equation(2)

$$\begin{cases} \dot{p} = (c_1 r + c_2 p)q + c_3 M_x + c_4 M_z \\ \dot{q} = c_5 pr - c_6(p^2 - r^2) + c_7 M_y \\ \dot{r} = (c_8 p - c_2 r)q + c_4 M_x + c_9 N_z \end{cases}$$

equation(3)

$$\begin{cases} \dot{V}_x = rV_y - qV_z - g\sin\theta + F_x/m \\ \dot{V}_y = -rV_x + pV_z + g\sin\phi\cos\theta + F_y/m \\ \dot{V}_z = qV_x - pV_y + g\cos\phi\cos\theta + F_z/m \end{cases}$$

equation(4)

$$\begin{cases} \dot{\phi} = p + (r\cos\phi + q\sin\phi)\tan\theta \\ \dot{\theta} = q\cos\phi - r\sin\phi \\ \dot{\psi} = (r\cos\phi + q\sin\phi)/\cos\theta \end{cases}$$

equation(5)

$$\begin{cases} \dot{x}_g = V_x \cos\theta \cos\psi + V_y(\sin\phi \sin\theta \cos\psi - \cos\phi \sin\psi) \\ \qquad + V_z(\sin\phi \sin\psi + \cos\phi \sin\theta \cos\psi) \\ \dot{y}_g = V_x \cos\theta \sin\psi + V_y(\sin\phi \sin\theta \sin\psi + \cos\phi \cos\psi) \\ \qquad + V_z(-\sin\phi \cos\psi + \cos\phi \sin\theta \sin\psi) \\ \dot{h} = V_x \sin\theta - V_y \sin\phi \cos\theta - V_z \cos\phi \cos\theta \end{cases}$$

where *p*, *q* and *r* are the roll, pitch, and yaw angular rates; *Vx*, *Vy* and *Vz* the components of flight velocity along the body axes; ϕ, θ and ψ the roll, pitch, and yaw attitude angles; x_g, y_g and *h* are respectively the north, east, and vertical components of the aircraft position in the locally-level geographic frame on the surface of the Earth;*m* is the mass of the aircraft, and *g* the gravity acceleration; the constants *ci*($i = 1, 2, \ldots, 9$) are defined by

equation(6)

$$\begin{cases} c_1 = [(J_y - J_z)J_z - J_{xz}^2]/\Gamma \\ c_2 = (J_x - J_y + J_z)/\Gamma; \quad c_3 = J_z/\Gamma \\ c_4 = J_{xz}/\Gamma; \quad c_5 = (J_z - J_x)/J_y; \quad c_6 = J_{xz}/J_y \\ c_7 = 1/J_y; \quad c_8 = [J_x(J_x - J_y) + J_{xz}^2]/\Gamma \\ c_9 = J_x/\Gamma; \quad \Gamma = J_xJ_z - J_{xz}^2 \end{cases}$$

where *Ji* ($i = x, y, z$) is the moment of inertia about *i* axis, and *Jxz* the cross-product of inertia; *Mx*, *My* and *Mz* are the roll, pitch, and yaw moments, which can be described as

equation(7)

$$\begin{cases} M_x = \rho V^2 S b C_l / 2 \\ M_y = \rho V^2 S \bar{c} C_m / 2 \\ M_z = \rho V^2 S b C_n / 2 \end{cases}$$

where ρ is the air density, V the flight velocity, S the wing reference area, b the wing span, and $\bar{c}$ the wing mean geometric chord; Cl, Cm and Cn are the roll, pitch and yaw moment coefficients.

In Eq. (3), Fx, Fy and Fz are the components of the resultant force including aerodynamic force and engine thrust along the body axes, and they are defined by

equation(8)

$$\begin{cases} F_x = P + L \sin\alpha - Y \cos\alpha \sin\beta - D \cos\alpha \cos\beta \\ F_y = Y \cos\beta - D \sin\beta \\ F_z = -L \cos\alpha - Y \sin\alpha \sin\beta - D \sin\alpha \cos\beta \end{cases}$$

where α and β are the angles of attack and sideslip; the engine thrust $P = C_P \delta_P$, whereC_P is the thrust coefficient, and δ_P the throttle setting; D, L and Y are the drag, lift and side-force, which can be described as

equation(9)

$$\begin{cases} D = \rho V^2 S C_D / 2 \\ L = \rho V^2 S C_L / 2 \\ Y = \rho V^2 S C_Y / 2 \end{cases}$$

where *CD*, *CL* and *CY* are the drag, lift and side-force coefficients.

The expressions of $\boldsymbol{M}_0$ and $\Delta\boldsymbol{M}$ in Fig. 1 will be given in the following text. According to Eq. (1), the roll, pitch, and yaw moments *Mx*, *My* and *Mz* can be respectively separated into two parts as follows:

equation(10)

$$\begin{cases} M_x = M_x^0 + \Delta M_x \\ M_y = M_y^0 + \Delta M_y \\ M_z = M_z^0 + \Delta M_z \end{cases}$$

where Δ*Mx*, Δ*My* and Δ*Mz* are the moment increments produced by control surface deflections, and $\boldsymbol{M}_0 = \begin{bmatrix} M_x^0 & M_y^0 & M_z^0 \end{bmatrix}^{\mathrm{T}}$ is given by

equation(11)

$$\begin{cases} M_x^0 = \rho V^2 Sb(C_{l\beta}\beta + C_{lp}\bar{p} + C_{lr}\bar{r})/2 \\ M_y^0 = \rho V^2 S\bar{c}(C_{m,\alpha=0} + C_{m\alpha}\alpha + C_{mq}\bar{q} + C_{m\dot{\alpha}}\bar{\dot{\alpha}})/2 \\ M_z^0 = \rho V^2 Sb(C_{n\beta}\beta + C_{np}\bar{p} + C_{nr}\bar{r})/2 \end{cases}$$

where $C_{(*)}$ is the aerodynamic derivatives; $\bar{p}$, $\bar{q}$, $\bar{r}$ and $\bar{\dot{\alpha}}$ are defined by

equation(12)

$$\begin{cases} \bar{p} = pb/(2V), & \bar{r} = rb/(2V) \\ \bar{q} = q\bar{c}/(2V), & \bar{\dot{\alpha}} = \dot{\alpha}\bar{c}/(2V) \end{cases}$$

and $\dot{\alpha}$ is the derivative of the angle of attack.

Substituting Eqs. (10), (11) and (12) into Eq. (2), Eq. (2) can be rewritten in the affine nonlinear form

equation(13)

$$\begin{bmatrix} \dot{p} \\ \dot{q} \\ \dot{r} \end{bmatrix} = \begin{bmatrix} f_p \\ f_q \\ f_r \end{bmatrix} + \boldsymbol{B}_c \begin{bmatrix} \Delta M_x \\ \Delta M_y \\ \Delta M_z \end{bmatrix}$$

where *fq*, *fq*, *fr* and $\boldsymbol{B}_c$ are

equation(14)

$$\begin{cases} f_p = (c_1 r + c_2 p)q + c_3 M_x + c_4 M_z \\ f_q = c_5 pr - c_6(p^2 - r^2) + c_7 M_y \\ f_r = (c_8 p - c_2 r)q + c_4 M_x + c_9 M_z \\ \boldsymbol{B}_c = \begin{bmatrix} c_3 & 0 & c_4 \\ 0 & c_7 & 0 \\ c_4 & 0 & c_9 \end{bmatrix} \end{cases}$$

We suppose $\Delta \boldsymbol{M} = [\Delta Mx\ \Delta My\ \Delta Mz]^T$, and according to Eq. (13), the actual moment increments $\Delta \boldsymbol{M}$ is derived by

equation(15)

$$\Delta \boldsymbol{M} = \boldsymbol{B}_c^{-1} [\dot{p} - f_p \quad \dot{q} - f_q \quad \dot{r} - f_r]^T$$

where $\dot{p}$, $\dot{q}$ and $\dot{r}$ are measured by the angular acceleration sensors.

2.2. Dynamic inversion attitude controller design

Having separated the attitude states into fast and slow dynamics, the feedback is used to provide the system with desirable dynamics. The fast dynamics of the aircraft attitude has been given in Eq. (2). And the desired attitude angular rates dynamics is specified by[1]

equation(16)

$$\begin{bmatrix} \dot{p}_d \\ \dot{q}_d \\ \dot{r}_d \end{bmatrix} = \begin{bmatrix} \omega_p & 0 & 0 \\ 0 & \omega_q & 0 \\ 0 & 0 & \omega_r \end{bmatrix} \begin{bmatrix} p_c - p \\ q_c - q \\ r_c - r \end{bmatrix}$$

where ωp, ωq and ωr are the design parameters; p_c, q_c, and r_c the commanded angular rates given by the slow dynamics loop.

Replacing the $[\dot{p} \quad \dot{q} \quad \dot{r}]^T$ in the left of Eq. (13) by $[\dot{p}_d \quad \dot{q}_d \quad \dot{r}_d]^T$ from Eq. (15), we can derive the desired moment increment $\Delta \boldsymbol{M}_d$:

equation(17)

$$\Delta \boldsymbol{M}_d = \boldsymbol{B}_c^{-1} [\dot{p}_d - f_p \quad \dot{q}_d - f_q \quad \dot{r}_d - f_r]^T$$

The slow dynamics of the aircraft attitude has been given in Eq. (4). And the desired attitude angular dynamics is similar to Eq. (16):

equation(18)

$$\begin{bmatrix} \dot{\phi}_d \\ \dot{\theta}_d \\ \dot{\psi}_d \end{bmatrix} = \begin{bmatrix} \omega_\phi & 0 & 0 \\ 0 & \omega_\theta & 0 \\ 0 & 0 & \omega_\psi \end{bmatrix} \begin{bmatrix} \phi_c - \phi \\ \theta_c - \theta \\ \psi_c - \psi \end{bmatrix}$$

where $\omega\phi$, $\omega\theta$ and $\omega\psi$ are the design parameters; ϕ_c, θ_c, and ψ_c the commanded attitude angles given by designer.

Replacing $[\dot{\phi} \quad \dot{\theta} \quad \dot{\psi}]^T$ in the left of Eq. (4) by $[\dot{\phi}_d \quad \dot{\theta}_d \quad \dot{\psi}_d]^T$ from Eq. (18), we can derive the commanded angular rates $[p_c \quad q_c \quad r_c]^T$ as follows:

equation(19)

$$\begin{bmatrix} p_c \\ q_c \\ r_c \end{bmatrix} = \begin{bmatrix} 1 & \tan\theta\sin\phi & \tan\theta\cos\phi \\ 0 & \cos\phi & -\sin\phi \\ 0 & \sin\phi/\cos\theta & \cos\phi/\cos\theta \end{bmatrix}^{-1} \begin{bmatrix} \dot{\phi}_d \\ \dot{\theta}_d \\ \dot{\psi}_d \end{bmatrix}$$

2.3. Control allocation problem

The objective of control allocation is to determined the actual control vector $\boldsymbol{u} \in \mathbf{R}m$according to the virtual control vector $\boldsymbol{v} \in \mathbf{R}l$, where $m > l$.

Mathematically, given $\boldsymbol{v}(t)$, then $\boldsymbol{u}(t)$ is derived by

equation(20)

$$f(\boldsymbol{u}(t))=\boldsymbol{v}(t)$$

where $f: \mathbf{R}m \mapsto \mathbf{R}l$ is the nonlinear mapping from $\boldsymbol{u}(t)$ to $\mathbf{v}(t)$. The actuator dynamics can be described as

equation(21)

$$\dot{\boldsymbol{\delta}} = g(\boldsymbol{\delta}, \boldsymbol{u})$$

where $\boldsymbol{\delta} \in \mathbf{R}m$ is the actuator deflection, and Eq. (21) is subject to a set of constraints

equation(22)

$$\boldsymbol{\delta}_{\min} \leqslant \boldsymbol{\delta} \leqslant \boldsymbol{\delta}_{\max}, \quad |\dot{\boldsymbol{\delta}}| \leqslant \dot{\boldsymbol{\delta}}_{\max}$$

where $\boldsymbol{\delta}_{\min}$, $\boldsymbol{\delta}_{\max}$ and $\dot{\boldsymbol{\delta}}_{\max}$ are the lower and upper actuator position and rate constraints, respectively. Due to the typically fast actuator dynamics, we assume $\boldsymbol{\delta} = \boldsymbol{u}$.

In a digital flight control system, it is reasonable to transform the actuator constraints from Eq. (22) to the following formula[16]:

equation(23)

$$\begin{cases} \underline{\boldsymbol{u}}(t) \leqslant \boldsymbol{u}(t) \leqslant \bar{\boldsymbol{u}}(t) \\ \underline{\boldsymbol{u}}(t) = \max(\boldsymbol{\delta}_{\min}, (\boldsymbol{u}(t-T) - \dot{\boldsymbol{\delta}}_{\max} T)) \\ \bar{\boldsymbol{u}}(t) = \min(\boldsymbol{\delta}_{\max}, (\boldsymbol{u}(t-T) + \dot{\boldsymbol{\delta}}_{\max} T)) \end{cases}$$

where T is the sample time.

2.4. Analytical solution of closed-loop dynamic control allocation

The dynamic control allocation can be expressed as the following sequential quadratic-programming problem if no actuators are saturated.

equation(24)

$$\begin{cases} \boldsymbol{u}_v(t) = \underset{\boldsymbol{u}_v(t)\in\boldsymbol{\Omega}}{\arg\min}\{\|\boldsymbol{W}_1(\boldsymbol{u}_v(t)-\boldsymbol{u}_s(t))\|^2 + \|\boldsymbol{W}_2(\boldsymbol{u}_v(t)-\boldsymbol{u}_v(t-T))\|^2\} \\ \boldsymbol{\Omega} = \underset{\underline{\boldsymbol{u}}(t)\leqslant \boldsymbol{u}_v(t)\leqslant \bar{\boldsymbol{u}}(t)}{\arg\min} \|\boldsymbol{W}_{\Omega}(\boldsymbol{B}\boldsymbol{u}_v(t)-\boldsymbol{v}(t))\| \end{cases}$$

where $\boldsymbol{u}_s(t) \in \mathbf{R}m$ is the desired steady-state control input; $\boldsymbol{W}_1$, $\boldsymbol{W}_2$ and $\boldsymbol{W}\Omega$ are the square matrices of proper dimensions; $\|\boldsymbol{x}\|$ denotes the Euclidean norm defined by $\|\boldsymbol{x}\| = \sqrt{\boldsymbol{x}^{\mathrm{T}}\boldsymbol{x}}$.

LEMMA 1.

(Ref.[12])*Let us remove the constraints of actuators, then the closed form solution to Eq.*(24)*is*

equation(25)

$$\boldsymbol{uv}(\mathrm{k}) = \boldsymbol{E}\boldsymbol{u}_s(\mathrm{k}) + \boldsymbol{F}\boldsymbol{uv}(\mathrm{k}-1) + \boldsymbol{G}\boldsymbol{v}(\mathrm{k})$$

Where

equation(26)

$$\begin{cases} E = (I - GB)W^{-2}W_1^2 \\ F = (I - GB)W^{-2}W_2^2 \\ G = W^{-1}(BW^{-1})^{\dagger} \\ W = \sqrt{W_1^2 + W_2^2} \end{cases}$$

Where $\boldsymbol{I}$*is the identity matrix, and the symbol "†" denotes the pseudoinverse operator defined as*$\boldsymbol{A}^{\dagger} = \boldsymbol{A}^{T}(\boldsymbol{AA}T)^{-1}$.

In the steady-state, we assume the error $\boldsymbol{v}(k) = \Delta \boldsymbol{M}_d(k) - \Delta \boldsymbol{M}(k)$ is zero, therefore, the desired control input $\boldsymbol{u}_s(k) = \boldsymbol{0}$, and $\boldsymbol{uv}(k) = \boldsymbol{uv}(k - 1)$.

Then, Eq. (25) reduces to

equation(27)

$$\boldsymbol{uv}(k) = (\boldsymbol{I} - \boldsymbol{F})^{-1}\boldsymbol{G}\boldsymbol{v}(k)$$

The input of actuator $\boldsymbol{u}(k)$ in Fig. 1 is

equation(28)

$$\boldsymbol{u}(k) = \boldsymbol{u}(k-1) + \boldsymbol{uv}(k)$$

3. STABILITY FOR CLOSED-LOOP DYNAMIC CONTROL ALLOCATION

The stability for the closed-loop dynamic control allocation system is of significant importance. In order to validate the stability, the input-output relation of closed-loop dynamic control allocation is given in the following text.

The actual moment increment $\Delta\boldsymbol{M}(k)$ can be derived by

equation(29)

$$\Delta\boldsymbol{M}(\mathrm{k})=\boldsymbol{B}_{\mathrm{r}}(\mathrm{k})\boldsymbol{u}(\mathrm{k})$$

where $\boldsymbol{B}_{\mathrm{r}}(k)$ is the actual control effectiveness matrix, which is changed with different flight conditions.

Substituting Eqs. (27) and (28) into Eq. (29), we have

equation(30)

$$\Delta\boldsymbol{M}(\mathrm{k})=\boldsymbol{B}_{\mathrm{r}}(\mathrm{k})\boldsymbol{u}(\mathrm{k}-1)+\boldsymbol{B}_{\mathrm{r}}(\mathrm{k})(\boldsymbol{I}-\boldsymbol{F})^{-1}\boldsymbol{G}\boldsymbol{v}(\mathrm{k})$$

where the input of dynamic control allocation $\boldsymbol{v}(k)$ is given by

equation(31)

$$\boldsymbol{v}(\mathrm{k})=\Delta\boldsymbol{M}_{\mathrm{d}}(\mathrm{k})-\Delta\boldsymbol{M}(\mathrm{k})$$

The change of $\boldsymbol{B}_r(k)$ is small during the sample time, we obtain

equation(32)

$$\boldsymbol{B}_r(k)=\boldsymbol{B}_r(k-1)$$

Then substituting Eqs. (29), (31) and (32) into Eq. (30), we have

equation(33)

$$\Delta\boldsymbol{M}(k)=\Delta\boldsymbol{M}(k-1)+\boldsymbol{V}(k)(\Delta\boldsymbol{M}_d(k)-\Delta\boldsymbol{M}(k))$$

where

equation(34)

$$\boldsymbol{V}(k)=\boldsymbol{B}_r(k)(\boldsymbol{I}-\boldsymbol{F})^{-1}\boldsymbol{G}, \quad \boldsymbol{V}\in \mathbf{R}^{l\times l}$$

The input–output relation of closed-loop dynamic control allocation is obtained according to the *z* transform of Eq. (33):

equation(35)

$$\frac{\Delta\boldsymbol{M}(z)}{\Delta\boldsymbol{M}_d(z)}=\frac{z\boldsymbol{V}(z)}{z(\boldsymbol{I}+\boldsymbol{V}(z))-\boldsymbol{I}}$$

3.1. Stability in the absence of actuator failures

In this part, we consider the stability for closed-loop dynamic allocation when the actuators are healthy.

LEMMA 2 SCHUR LEMMA.

Given $\boldsymbol{A} \in \mathbf{R}^{n \times n}$ *with eigenvalues* $\lambda_1, \lambda_2, \ldots, \lambda_n$ *in any prescribed order, there is an orthogonal matrix* $\boldsymbol{P} \in \boldsymbol{E}^{n \times n}$*, where* $\boldsymbol{E}$ *is the set of orthogonal matrix. And we have*

$$\boldsymbol{P}^{-1}\boldsymbol{A}\boldsymbol{P} = \boldsymbol{B} = \begin{bmatrix} b_{11} & b_{12} & \cdots & b_{1n} \\ 0 & b_{22} & \cdots & b_{2n} \\ \vdots & \ddots & \ddots & \vdots \\ 0 & \cdots & 0 & b_{nn} \end{bmatrix}$$

is upper triangular, with diagonal entries $b_{ii} = \lambda_i$ $(i = 1, 2, \ldots, n)$.

THEOREM 1.

Let the closed form solution to Eq. (24) *be defined in Lemma* 1*, and the closed-loop dynamic control allocation system is described as Eq.* (35)*. And* λ_i $(i = 1, 2, \ldots, n)$ *is the eigenvalue of* $\boldsymbol{V}(z)$*. The closed-loop dynamic control allocation is stable if* $|1 + \lambda_i| > 1$ *is satisfied.*

PROOF.

According to Eq. (35), the stability of the closed-loop allocation system is determined by the locations of the closed-loop poles or the roots of characteristic equation in the z plane. The characteristic equation is

equation(36)

$$z(\boldsymbol{I}+\boldsymbol{V}(z))-\boldsymbol{I}=\boldsymbol{0}$$

And the system is stable if any of the closed-loop characteristic roots lie inside the unit circle.

Considering Lemma 2, we obtain

equation(37)

$$\boldsymbol{V}(z)=\boldsymbol{P}\begin{bmatrix}\lambda_1 & b_{12} & \cdots & \cdots & b_{1n}\\ \boldsymbol{0} & \lambda_2 & b_{22} & \cdots & b_{2n}\\ \vdots & & \ddots & \ddots & \vdots\\ \vdots & & & \ddots & b_{n-1,n}\\ \boldsymbol{0} & 0 & \cdots & 0 & \lambda_n\end{bmatrix}\boldsymbol{P}^{-1}$$

Substituting Eq. (37) into Eq. (36), we have

equation(38)

$$z\boldsymbol{P}\boldsymbol{P}^{-1}+z\boldsymbol{P}\begin{bmatrix}\lambda_1 & b_{12} & \cdots & \cdots & b_{1n}\\ 0 & \lambda_2 & b_{22} & \cdots & b_{2n}\\ \vdots & & \ddots & \ddots & \vdots\\ \vdots & & & \ddots & b_{n-1,n}\\ 0 & 0 & \cdots & 0 & \lambda_n\end{bmatrix}\boldsymbol{P}^{-1}-\boldsymbol{P}\boldsymbol{P}^{-1}$$

$$=\boldsymbol{P}\left(z\boldsymbol{I}+z\begin{bmatrix}\lambda_1 & b_{12} & \cdots & \cdots & b_{1n}\\ 0 & \lambda_2 & b_{22} & \cdots & b_{2n}\\ \vdots & & \ddots & \ddots & \vdots\\ \vdots & & & \ddots & b_{n-1,n}\\ 0 & 0 & \cdots & 0 & \lambda_n\end{bmatrix}-\boldsymbol{I}\right)\boldsymbol{P}^{-1}=\boldsymbol{0}$$

The characteristic equation becomes

equation(39)

$$z_i + z_i\lambda_i - 1 = 0 \quad (i = 1, 2, \ldots, n)$$

The roots of characteristic equation are found to be

equation(40)

$$z_i = \frac{1}{1 + \lambda_i} \quad (i = 1, 2, \ldots, n)$$

If $|z_i| < 1$, then the system is stable and we obtain

equation(41)

$$|1 + \lambda_i| > 1$$

This completes the proof.

3.2. Stability in the presence of actuator failures

The stability condition for closed-loop control allocation is given in Theorem 1 when the actuators are healthy. However, the actuators of an aircraft can be affected by many types of failures in the flight missions. And there are two typical failures which are the loss of effectiveness and lock-in-place failures. Both of them are considered here.

ASSUMPTION 1.

Two actuator failures do not happen simultaneously.

The control signal $\boldsymbol{\delta}(t)$ (input to the plant, output of the actuator) can be described as the following formula to incorporate the actuator failures.

equation(42)

$$\begin{cases} \boldsymbol{\delta}(t) = (\boldsymbol{\Xi} + \boldsymbol{\Phi}_u)\boldsymbol{u}(t) + \boldsymbol{\Phi}_w \boldsymbol{w}(t) \\ \boldsymbol{\Xi} = \mathrm{diag}(d_1^e, d_2^e, \cdots, d_m^e), \quad 0 < d_i^e \leqslant 1 \\ \boldsymbol{\Phi}_u = \mathrm{diag}(d_1^u, d_2^u, \cdots, d_m^u), \quad d_i^u = \{0, 1\} \\ \boldsymbol{\Phi}_w = \mathrm{diag}(d_1^w, d_2^w, \cdots, d_m^w), \quad d_i^w = \{0, 1\} \\ \boldsymbol{w} = [w_1 \quad w_2 \quad \cdots \quad w_m]^{\mathrm{T}}, \quad \delta_{\min}^i \leqslant w_i \leqslant \delta_{\max}^i \end{cases}$$

where $\delta_{\min}^i$ and $\delta_{\max}^i$ present the i th element of $\boldsymbol{\delta}_{\min}$ and $\boldsymbol{\delta}_{\max}$ in Eq. (22), respectively. And the actuator i has a loss of effectiveness failure if $0 < d_i^e < 1$, $d_j^e = 1$, $j \neq i$ and $\boldsymbol{\Phi}_u = \boldsymbol{\Phi}_w = \mathbf{0}$; the actuator i is stuck at the degree w_i if $d_i^u = 0$, $d_j^u = 1$, $d_i^w = 1$, $d_j^w = 0$, $j \neq i$ and $\boldsymbol{\Xi} = \mathbf{0}$.

In the presence of the actuator loss of effectiveness failure, the actual moment increment $\Delta\boldsymbol{M}(k)$ will be changed. Considering Eqs. (29) and (42), we have

equation(43)

$$\Delta\boldsymbol{M}(\mathrm{k}) = \boldsymbol{B}_{\mathrm{r}}(\mathrm{k})\boldsymbol{\Xi}\boldsymbol{u}(\mathrm{k}) = \boldsymbol{B}_{\mathrm{f}}(\mathrm{k})\boldsymbol{u}(\mathrm{k})$$

We can see from Eq. (43) the actuator failures change the control effectiveness matrix. And the changed control effectiveness matrix $\boldsymbol{B}_{\mathrm{f}}$ can be written as

equation(44)

$$\boldsymbol{B}_{\mathrm{f}}(k) = \boldsymbol{B}_{\mathrm{r}}^{0}(k) + \Delta\boldsymbol{B}_{\mathrm{r}}(k)$$

where $\boldsymbol{B}_{\mathrm{r}}^{0}(k)$ is the control effectiveness matrix without failures, and $\Delta\boldsymbol{B}_{\mathrm{r}}(k)$ the control effectiveness matrix change produced by the actuator loss of effectiveness failure. Substituting Eq. (44) into Eq. (34), we have

equation(45)

$$\boldsymbol{V}(k) = \left(\boldsymbol{B}_{\mathrm{r}}^{0}(k) + \Delta\boldsymbol{B}_{\mathrm{r}}(k)\right)(\boldsymbol{I} - \boldsymbol{F})^{-1}\boldsymbol{G} = \boldsymbol{V}_{0}(k) + \Delta\boldsymbol{V}(k)$$

where $\boldsymbol{V}_{0}(k) = \boldsymbol{B}_{\mathrm{r}}^{0}(k)(\boldsymbol{I} - \boldsymbol{F})^{-1}\boldsymbol{G}$ is the matrix without failures, and the error caused by the actuator loss of effectiveness failure is

equation(46)

$$\Delta\boldsymbol{V}(\mathrm{k})=\Delta\boldsymbol{B}_{\mathrm{r}}(\mathrm{k})(\boldsymbol{I}-\boldsymbol{F})^{-1}\boldsymbol{G}$$

ASSUMPTION 2.

The $V_0(k)$ and $\Delta V(k\ )$ commute, and have eigenvalues $\lambda_1^0, \lambda_2^0, \ldots, \lambda_n^0$ and $\lambda_1^\Delta, \lambda_2^\Delta, \ldots, \lambda_n^\Delta$, respectively.

COROLLARY 1.

The actuator loss of effectiveness failure happens. Considering Eqs. (35) and (45)*andAssumption 2, if the eigenvalues* $\lambda_i^\Delta\ (i = 1, 2, \ldots, n)$ *of the error matrix* $\Delta V(k)$ *and the eigenvalues* $\lambda_i^0\ (i = 1, 2, \ldots, n)$ *of the matrix* $V0(k)$ *satisfy the following formula:*

equation(47)

$$\lambda_i^\Delta > -\lambda_i^0 \quad \text{or} \quad \lambda_i^\Delta < -(\lambda_i^0 + 2)$$

then the closed-loop dynamic control allocation is stable in the presence of the actuator loss of effectiveness failure.

PROOF.

According to Assumption 2, the eigenvalues of $V_0(k) + \Delta V(k\ )$ are $\lambda_i^0 + \lambda_i^\Delta\ (i = 1, 2, \ldots, n)$.

The actual control effectiveness matrix changes in the presence of the actuator loss of effectiveness failure; however, the stability of the closed-loop system is desired.

Considering Theorem 1, we obtain

equation(48)

$$|\lambda_i^\Delta + \lambda_i^0 + 1| > 1 \Rightarrow \lambda_i^\Delta > -\lambda_i^0 \quad \text{or} \quad \lambda_i^\Delta < -(\lambda_i^0 + 2)$$

This completes the proof.

In the presence of the actuator lock-in-place failure, the actuator *i* is stuck at the degree*wi*. It is equivalent to adding the disturbance to the actual moment increment $\Delta\boldsymbol{M}(k)$. Considering Eqs. (29) and (42), we have

equation(49)

$$\Delta\boldsymbol{M}(\mathrm{k})=\boldsymbol{B}_{\mathrm{r}}(\mathrm{k})(\boldsymbol{\Phi u u}(\mathrm{k})+\boldsymbol{\Phi w w}(\mathrm{k}))$$

COROLLARY 2.

The actuator lock-in-place failure happens. If the eigenvalues λ_i^u *of the matrix* $\boldsymbol{Vu}(k)$*satisfy the following formula:*

equation(50)

$$\left|1+\lambda_i^u\right|>1 \quad (i=1,2,\ldots,n)$$

where

equation(51)

$$\boldsymbol{Vu}(\mathrm{k})=\boldsymbol{B}_{\mathrm{r}}(\mathrm{k})\boldsymbol{\Phi u}(\boldsymbol{I}-\boldsymbol{F})^{-1}\boldsymbol{G}$$

and the steady-state inputs of actuator $\boldsymbol{u}$ (∞) *satisfy*

equation(52)

$$\boldsymbol{B}_{\mathrm{r}}(\infty)(\boldsymbol{\Phi}_u \boldsymbol{u}(\infty) + \boldsymbol{\Phi}_w \boldsymbol{w}(\infty)) = \boldsymbol{0}$$
$$(\boldsymbol{\delta}_{\min} \leqslant \boldsymbol{u}(\infty), \boldsymbol{w}(\infty) \leqslant \boldsymbol{\delta}_{\max})$$

then the closed-loop dynamic control allocation is stable in the presence of the actuator lock-in-place failure.

PROOF.

Substituting Eqs. (27) and (28) into Eq. (49), we have

equation(53)

$$\Delta \boldsymbol{M}(\mathrm{k}) = \boldsymbol{B}_{\mathrm{r}}(\mathrm{k})[\boldsymbol{\Phi u}(\boldsymbol{u}(\mathrm{k}-1) + (\boldsymbol{I}-\boldsymbol{F})^{-1}\boldsymbol{G}\boldsymbol{v}(\mathrm{k})) + \boldsymbol{\Phi w}\boldsymbol{w}(\mathrm{k})]$$

We derive $\boldsymbol{w}(k) = \boldsymbol{w}(k-1)$ because the actuator is stuck. And according to Eqs. (32),(33) and (53), we have

equation(54)

$$\Delta \boldsymbol{M}(\mathrm{k}) = \Delta \boldsymbol{M}(\mathrm{k}-1) + \boldsymbol{Vu}(\mathrm{k})(\Delta \boldsymbol{M}_{\mathrm{d}}(\mathrm{k}) - \Delta \boldsymbol{M}(\mathrm{k}))$$

where the matrix $\boldsymbol{Vu}(k)$ is given in Eq. (51).

Considering Theorem 1, we obtain $|1 + \lambda_i^u| > 1$.

Otherwise, the steady-state moment increment $\Delta \boldsymbol{M}(\infty) = 0$. According to Eq. (49), we have

equation(55)

$$\boldsymbol{B}_{\mathrm{r}}(\infty)(\boldsymbol{\Phi u u}(\infty)+\boldsymbol{\Phi w w}(\infty))=\mathbf{0}$$

and $\boldsymbol{u}(\infty)$ and $\boldsymbol{w}(\infty)$ must satisfy the following position limits of the actuators:

equation(56)

$$\boldsymbol{\delta}_{\min}\leqslant\boldsymbol{u}(\infty),\boldsymbol{w}(\infty)\leqslant\boldsymbol{\delta}_{\max}$$

which completes the proof.

4. SIMULATION RESULTS

In this section, we consider a canard rotor/wing (CRW) aircraft model (see Refs.[17] and [18]for more details) in the fixed-wing mode. The simulation model includes the moment, force, kinematic and navigation equations in Eqs. (2), (3), (4) and (5). And the specific aerodynamic force and moment coefficients of the CRW in the fixed-wing mode are given by

equation(57)

$$\begin{cases} C_L = C_{L0} + C_{L\alpha}\alpha + C_{L\delta_c}\delta_c + C_{L\delta_e}\delta_e \\ C_D = C_{D1}C_L^2 + C_{D2}C_L + C_{D3} \\ C_Y = C_{Y\beta}\beta + C_{Y\delta_{rL}}\delta_{rL} + C_{Y\delta_{rR}}\delta_{rR} \end{cases}$$

equation(58)

$$\begin{cases} C_l = C_{l\delta_{aL}}\delta_{aL} + C_{l\delta_{aR}}\delta_{aR} + C_{l\delta_{rL}}\delta_{rL} + C_{l\delta_{rR}}\delta_{rR} \\ \qquad + C_{l\beta}\beta + C_{lp}\bar{p} + C_{lr}\bar{r} \\ C_m = C_{m1}C_L^2 + C_{m2}C_L + C_{m3} + C_{m\delta_e}\delta_e + C_{m\delta_e}\delta_e \\ \qquad + C_{mq}\bar{q} + C_{m\dot{\alpha}}\bar{\dot{\alpha}} \\ C_n = C_{n\delta_{aL}}\delta_{aL} + C_{n\delta_{aR}}\delta_{aR} + C_{n\delta_{rL}}\delta_{rL} + C_{n\delta_{rR}}\delta_{rR} \\ \qquad + C_{n\beta}\beta + C_{np}\bar{p} + C_{nr}\bar{r} \end{cases}$$

where α,β and $\bar{\dot{\alpha}}$ can be derived by

equation(59)

$$\begin{cases} \alpha = \arctan(V_z/V_x) \\ \beta = \arcsin(V_y/V) \\ \dot{\alpha} = [(V_x\dot{V}_z - V_z\dot{V}_x)\cos\alpha]/V_x^2 \end{cases}$$

where the flight velocity $V = \sqrt{V_x^2 + V_y^2 + V_z^2}$. In addition, we can derive $\bar{\dot{\alpha}}$ according to Eqs. (12) and (59). In addition, the thrust coefficient of this aircraft in the fixed-wing mode is C_P = 2487.

The aerodynamic parameters are shown in Table 1, while the wing-planform and inertia parameters are shown in Table 2. These parameters are derived from both Ref.[19] and aerodynamic computing in terms of the aircraft geometry figuration.

Table 1. Aerodynamic parameters for the model.

Parameter	Value	Parameter	Value
C_{L0}	0.224	$C_{L\alpha}$	0.115
$C_{L\delta_c}$	0.0610	$C_{L\delta_e}$	0.0196
C_{D1}	0.0815	C_{D2}	-0.0465
C_{D3}	0.0756	$C_{Y\beta}$	-9.60×10^{-3}
$C_{Y\delta_{rL}}$	-1.45×10^{-3}	$C_{Y\delta_{rL}}$	-1.45×10^{-3}
$C_{l\delta_{aL}}$	2.50×10^{-4}	$C_{l\delta_{aR}}$	-2.50×10^{-4}
$C_{l\delta_{rL}}$	0	$C_{l\delta_{rR}}$	0
$C_{l\beta}$	-1.53×10^{-3}	$C_{l\bar{p}}$	-0.343
$C_{l\bar{r}}$	-5.54×10^{-3}	C_{m1}	-0.166
C_{m2}	0.147	C_{m3}	-0.182
$C_{m\delta_c}$	0.0790	$C_{m\delta_e}$	-0.0356
$C_{m\bar{q}}$	-57.1	$C_{m\dot{\alpha}}$	-11.6
$C_{n\delta_{aL}}$	9.00×10^{-6}	$C_{n\delta_{aR}}$	-9.00×10^{-6}
$C_{n\delta_{rL}}$	-9.65×10^{-4}	$C_{n\delta_{rR}}$	-9.65×10^{-4}
$C_{n\beta}$	-2.4×10^{-3}	$C_{n\bar{p}}$	8.46×10^{-3}
$C_{n\bar{r}}$	-0.560		

Table 2. Wing-planform and inertia parameters for the model.

Parameter	Value	Parameter	Value
b (m)	2.80	S (m^2)	1.21
$\bar{c}$ (m)	0.40	m (kg)	450
J_x(kg·m^2)	72.5	J_y(kg·m^2)	246.6
J_z(kg·m^2)	140.5	J_{xz}(kg·m^2)	0

The control inputs consist of the deflections for the left aileron δ_{aL}, the right aileron δ_{aR}, the left rudder δ_{rL} and the right rudder δ_{rR} in the lateral direction. And the canard δ_c and elevator δ_e with different frequency characteristics are available for pitch control in the longitudinal direction. Fig. 2 shows the configuration of the CRW aircraft.[20]

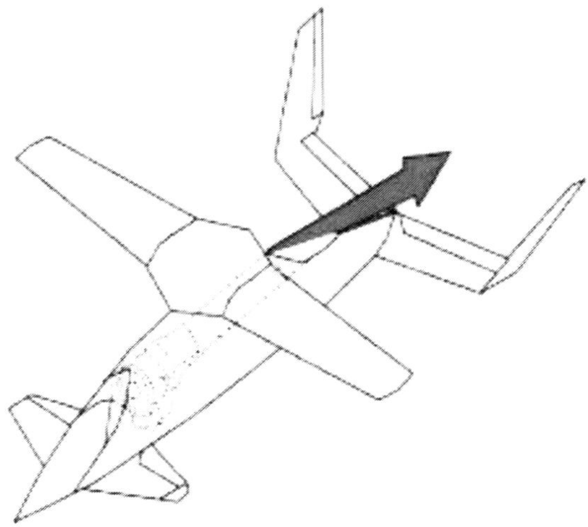

Figure. 2. CRW aircraft in the fixed wing mode.

The actuator dynamic is simplified to produce a simple first-order lag filter. The aileron, rudder, canard and elevator dynamics used in the simulation are described as

equation(60)

$$\begin{cases} \delta_{aL} = \frac{10}{s+10} u_{aL}, & \delta_{rL} = \frac{10}{s+10} u_{rL} \\ \delta_{aR} = \frac{10}{s+10} u_{aR}, & \delta_{rR} = \frac{10}{s+10} u_{rR} \\ \delta_{c} = \frac{20}{s+20} u_{c}, & \delta_{e} = \frac{10}{s+10} u_{e} \\ |\delta_{aL(R)}| \leqslant 25^{\circ}, \quad |\delta_{rL(R)}| \leqslant 25^{\circ}, & |\delta_{c}| \leqslant 15^{\circ}, \quad |\delta_{e}| \leqslant 25^{\circ} \end{cases}$$

where s denotes differential operator, $\delta_{aL(R)}$ denotes δ_{aL} or δ_{aR}, $\delta_{rL(R)}$ is similar to$\delta_{aL(R)}$, $(u_{aL}, u_{aR}, u_{rL}, u_{rR}, u_{c}, u_{e})$ and $(\delta_{aL}, \delta_{aR}, \delta_{rL}, \delta_{rR}, \delta_{c}, \delta_{e})$ are the input and output of the actuators, respectively. The canard actuator is faster than the elevator, whereas it is beneficial not to deflecting the canard at all to achieve low drag at trimmed flight. Thus, the canard is used to achieve the fast initial aircraft response, while the elevator is used solely at the trimmed flight. And we select the following constant matrix:

equation(61)

$$\begin{cases} \boldsymbol{W}_1 = \mathrm{diag}(0.5, 0.5, 0.1, 0.1, 60, 37) \\ \boldsymbol{W}_2 = \mathrm{diag}(0.4, 0.4, 0.8, 0.8, 0, 100) \end{cases}$$

The actual control effectiveness matrix without failures is given by

equation(62)

$$\boldsymbol{B}_{\mathrm{r}} = 10^{-4} \begin{bmatrix} 2.50 & -2.50 & 0 & 0 & 0 & 0 \\ 0.09 & -0.09 & -9.65 & -9.65 & 0 & 0 \\ 0 & 0 & 0 & 0 & 790 & -356 \end{bmatrix}$$

The design matrix is $\boldsymbol{B} = \boldsymbol{B}_{\mathrm{r}}$ whether the actuators are healthy or fail. However, the actual control effectiveness matrix $\boldsymbol{B}_{\mathrm{r}}$ changes when the actuator failures happen. The attitude controller parameters are $\omega\phi = 1.2$, $\omega\theta = 5$, $\omega\psi = 1.2$, $\omega p = 4.5$, $\omega q = 10$,$\omega r = 4.5$, and the sample time is $T = 0.02$ s.

In the following simulations, the initial altitude and velocity are 3000 m and 110 m/s, and the initial attitude angles and angular rates including $\alpha, \beta, \phi, \theta, \psi, p, q$ and r are zeros. In addition, the throttle setting $\delta_{\mathrm{P}} = 0.2$, while all the initial actuator deflections are zero.

4.1. No actuator failures

Considering Eqs. (26), (34), (61) and (62), the eigenvalues of the matrix $\boldsymbol{V}(k)$ are $\lambda_1 = \lambda_2 = \lambda_3 = 1$. According to Theorem 1, the designed closed-loop dynamic control allocation is stable. Fig. 3 and Fig. 4 show the attitude responses and actuator deflections using dynamic control allocation and closed-loop dynamic control allocation.

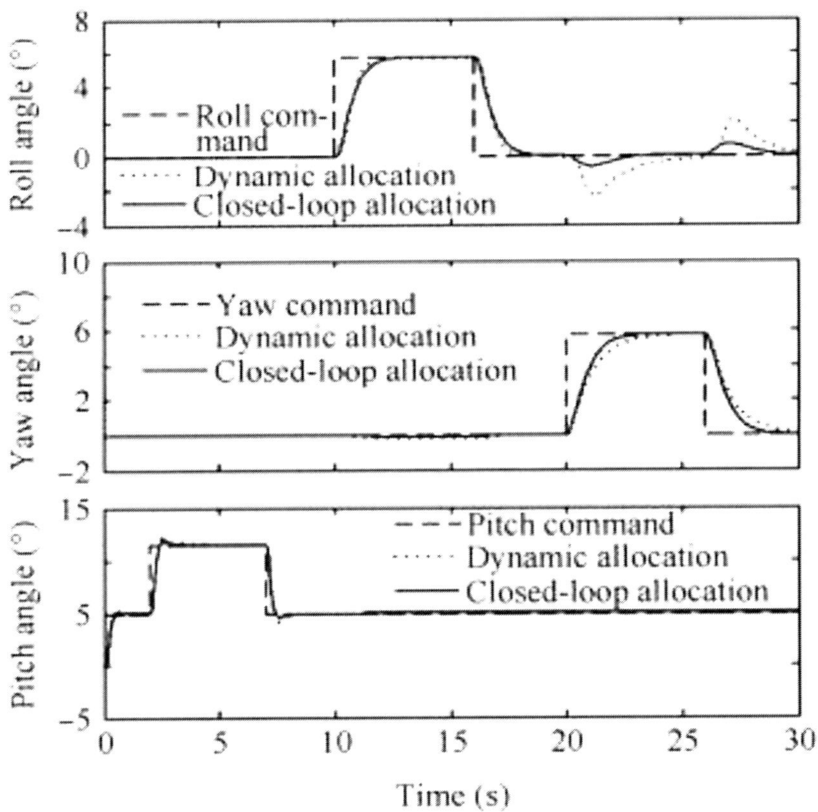

Figure. 3. Responses of attitude without failure.

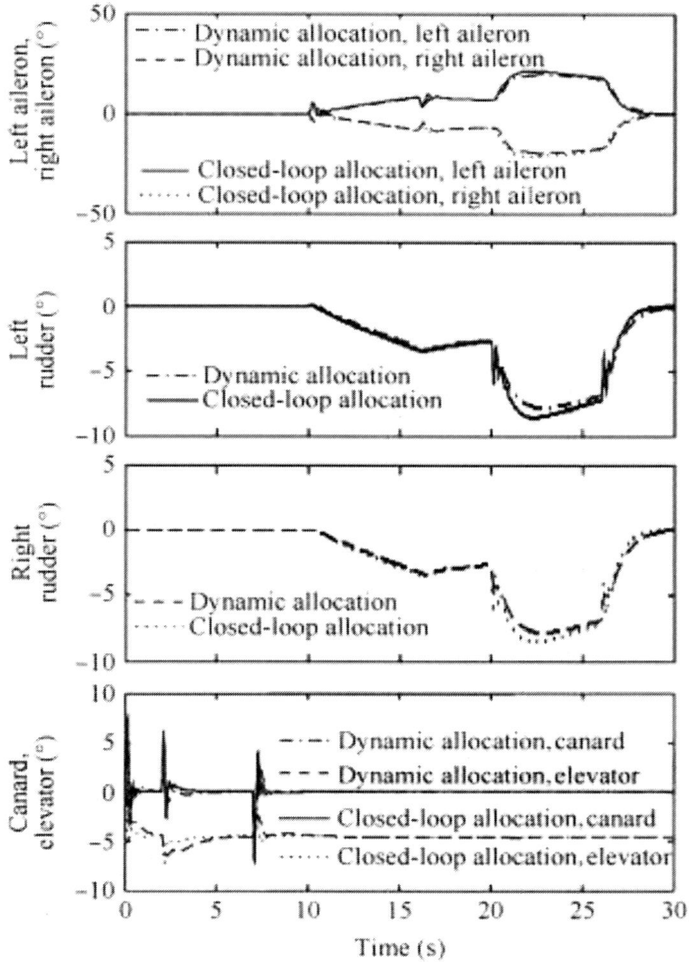

Figure. 4. Actuator deflections without failure.

The two methods have the similar attitude tracking performance and actuator responses when there are no actuator failures. And the canard deflections are zero in steady state for two methods in Fig. 4. However, there is shorter rise time in the yaw response, and there is smaller effect on the roll response when the yaw angle changes using the closed-loop dynamic control allocation method in Fig. 3. And there are smaller overshot and shorter settling time in the pitch response for the closed-loop dynamic control allocation.

4.2. Actuator failures

Two failure cases are considered; however, there is no fault detection subsystem to identify the actuator failures in this research. We suppose the longitudinal- and lateral-directional equations are decoupled, and there is only one actuator which has only one type failure in one simulation. Therefore, the simulations of the different types of actuator failures in longitudinal direction and those in lateral direction are performed respectively.

The actual control effectiveness matrices, $\boldsymbol{B}_{\mathrm{r\_lon}} \in \mathbf{R}^{1\times 2}$ and $\boldsymbol{B}_{\mathrm{r\_lat}} \in \mathbf{R}^{2\times 4}$, change while the design matrix, $\boldsymbol{B}_{\mathrm{lon}} \in \mathbf{R}^{1\times 2}$ and $\boldsymbol{B}_{\mathrm{lat}} \in \mathbf{R}^{2\times 4}$, maintain constant after the actuator failure happens. And the weighting $\boldsymbol{W}i_{\mathrm{\_lat}} \in \mathbf{R}^{4\times 4}$, $\boldsymbol{W}i_{\mathrm{\_lon}} \in \mathbf{R}^{2\times 2}$ (i = 1, 2) and design matrix are derived by Eqs. (61) and (62) using block matrix method.

equation(63)

$$\begin{cases} \boldsymbol{W}_1 = \mathrm{diag}(\boldsymbol{W}_{1\_\mathrm{lat}}, \boldsymbol{W}_{1\_\mathrm{lon}}) \\ \boldsymbol{W}_2 = \mathrm{diag}(\boldsymbol{W}_{2\_\mathrm{lat}}, \boldsymbol{W}_{2\_\mathrm{lon}}) \\ \boldsymbol{B}_{\mathrm{r}} = \begin{bmatrix} \boldsymbol{B}_{\mathrm{lat}} & \mathbf{0}_{2\times 2} \\ \mathbf{0}_{1\times 2} & \boldsymbol{B}_{\mathrm{lon}} \end{bmatrix} \end{cases}$$

4.2.1. Failure Case I: the loss of effectiveness failure

(1) Longitudinal actuator failure

The actuator failure is the loss of effectiveness failure of the elevator deflection in the pitch channel, while the canard is healthy. The elevator maintains only 50% of its effectiveness after 4 s, and we have the following formula according to Eq. (42).

equation(64)

$$\begin{bmatrix} \delta_c \\ \delta_e \end{bmatrix} = \begin{bmatrix} 1 & 0 \\ 0 & 0.5 \end{bmatrix} \begin{bmatrix} u_c \\ u_e \end{bmatrix} \quad (t \geqslant 4\ \text{s})$$

Both longitudinal attitude tracking and actuator deflections are shown in Fig. 5 and Fig. 6.

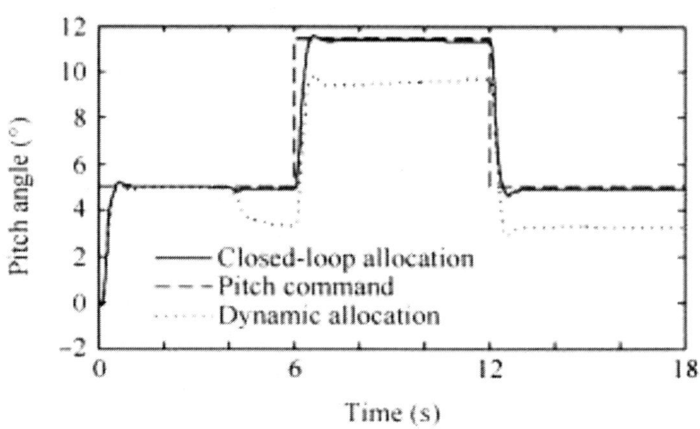

Figure. 5. Responses of pitch with elevator effectiveness loss.

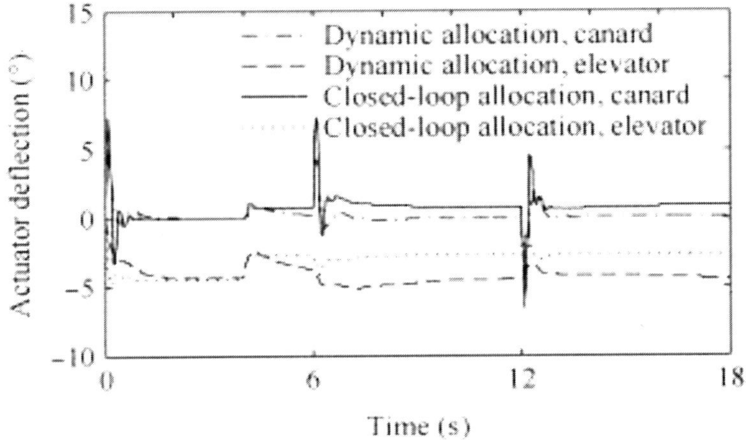

Figure. 6. Deflections of longitudinal actuators with elevator effectiveness loss.

Fig. 5 and Fig. 6 show that the dynamic control allocation method does not exactly distribute the desired pitch moment between canard and elevator after the failure happens. However, the closed-loop dynamic control allocation maintains the attitude tracking with little degradation, because there is a feedback from moment increment in the proposed approach, and the error between the desired and actual moment can be enforced to zero. To eliminate this error, the canard is used to compensate the elevator effectiveness loss. And the eigenvalues of $\boldsymbol{V}_0(k)$ and $\Delta \boldsymbol{V}(k\quad)$ are $\lambda_1^0 = 1$ and $\lambda_1^\Delta = -0.17$. According to Corollary 1, the closed-loop dynamic control allocation system is stable.

(2) Lateral actuator failure

The actuator failure is the loss of effectiveness failure of the right rudder deflection in the yaw channel, while the left rudder and ailerons are healthy. The right rudder maintains only 50% of its effectiveness after 12 s, and we have the following formula according to Eq. (42):

equation(65)

$$\begin{bmatrix} \delta_{aL} \\ \delta_{aR} \\ \delta_{rL} \\ \delta_{rR} \end{bmatrix} = \begin{bmatrix} 1 & & & \\ & 1 & & \\ & & 1 & \\ & & & 0.5 \end{bmatrix} \begin{bmatrix} u_{aL} \\ u_{aR} \\ u_{rL} \\ u_{rR} \end{bmatrix} \quad (t \geqslant 12\ \text{s})$$

Both lateral attitude tracking and actuator deflections are shown in Fig. 7 and Fig. 8.

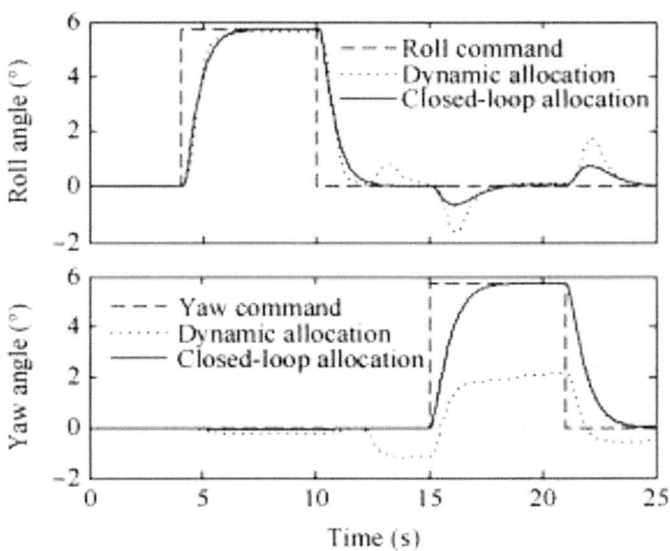

Figure. 7. Responses of lateral attitude with right rudder effectiveness loss.

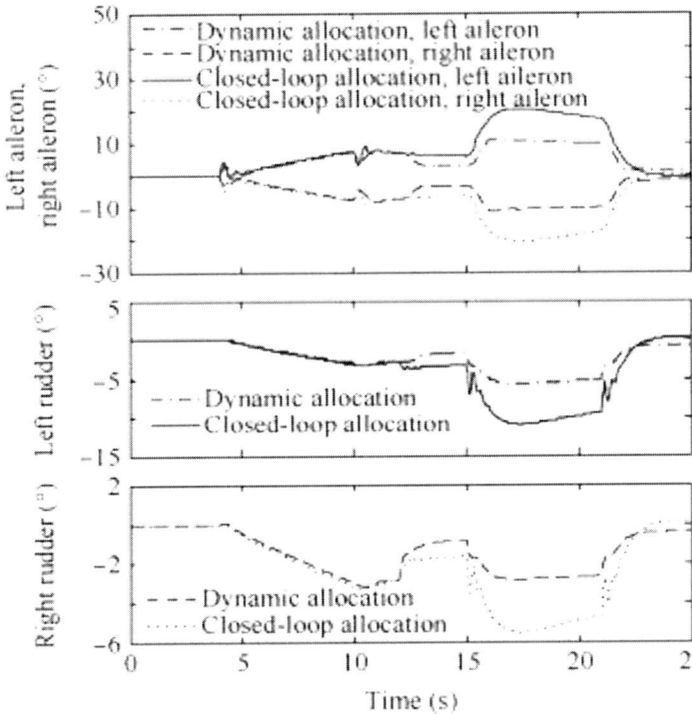

Figure. 8. Deflections of lateral actuators with right rudder effectiveness loss.

Fig. 7 shows that the desired lateral attitude commands, especially the yaw command, can be tracked by the proposed method after the failure happens. In Fig. 8, the deflections of all lateral actuators increase to derive the desired roll and yaw moments when the right rudder loses its effectiveness using the proposed method. And the eigenvalues of $\boldsymbol{V}_0(k)$ and $\Delta\boldsymbol{V}(k)$ are $\lambda_1^0 = 1$, $\lambda_1^\Delta = -0.25$, $\lambda_2^0 = 1$ and $\lambda_2^\Delta = 0$. According to Corollary 1, the closed-loop dynamic control allocation system is stable.

4.2.2. Failure Case II: lock-in-place failure

(1) Longitudinal actuator failure

The actuator failure is the lock-in-place failure of the elevator in the pitch channel, while the canard is healthy. The elevator is stuck at $-3°$ after 4 s, and we have the following formula according to Eq. (42).

equation(66)

$$\begin{bmatrix} \delta_c \\ \delta_e \end{bmatrix} = \begin{bmatrix} 1 & 0 \\ 0 & 0 \end{bmatrix} \begin{bmatrix} u_c \\ u_e \end{bmatrix} + \begin{bmatrix} 0 & 0 \\ 0 & 1 \end{bmatrix} \begin{bmatrix} 0 \\ -3 \end{bmatrix} \quad (t \geqslant 4\ \text{s})$$

Both longitudinal attitude tracking and actuator deflections are shown in Fig. 9 and Fig. 10.

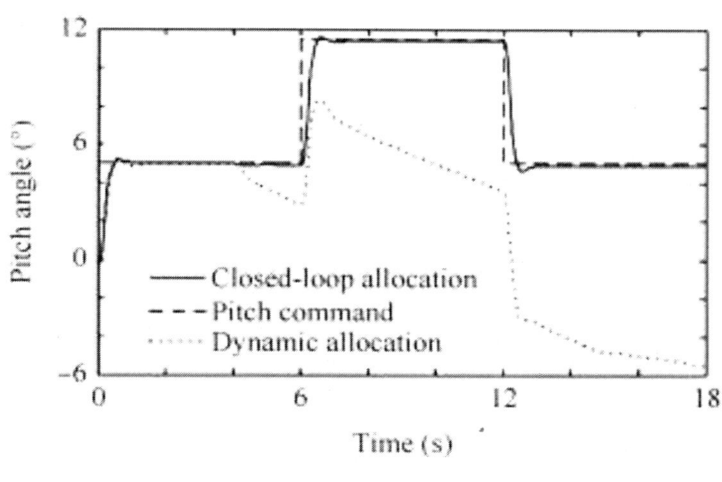

Figure. 9. Responses of pitch with elevator stuck.

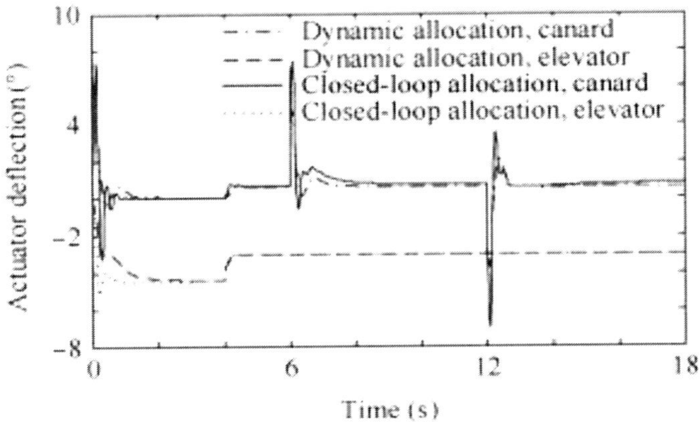

Figure. 10. Deflections of longitudinal actuators with elevator stuck.

Fig. 9 shows that the dynamic control allocation method does not track the pitch command after the elevator is stuck at −3°, and the attitude control system is unstable gradually. However, the proposed closed-loop dynamic control allocation maintains stability and attitude tracking after the failure happens. We can see the canard in the two methods deflects to counteract the disturbance from the locked elevator in Fig. 10. However, the magnitude of canard deflection is larger and the response is faster in the proposed method. And the existence of feedback loop forces the moment increments error to zero. The eigenvalue of $\boldsymbol{Vu}(k)$ is 0.65, and $\boldsymbol{u}(\infty) = [1.35° - 3°]^T$. According toCorollary 2, the closed-loop dynamic control allocation system is stable.

(2) Lateral actuator failure

The actuator failure is the lock-in-place failure of the left aileron in the roll channel, while the right aileron and rudders are healthy. The left aileron is stuck at 10° after 12 s, and we have the following equation according to Eq. (42):

equation(67)

$$\begin{bmatrix} \delta_{aL} \\ \delta_{aR} \\ \delta_{rL} \\ \delta_{rR} \end{bmatrix} = \begin{bmatrix} 0 & & & \\ & 1 & & \\ & & 1 & \\ & & & 1 \end{bmatrix} \begin{bmatrix} u_{aL} \\ u_{aR} \\ u_{rL} \\ u_{rR} \end{bmatrix} + \begin{bmatrix} 1 & & & \\ & 0 & & \\ & & 0 & \\ & & & 0 \end{bmatrix} \begin{bmatrix} 10 \\ 0 \\ 0 \\ 0 \end{bmatrix} \quad (t \geqslant 12\ \mathrm{s})$$

Both lateral attitude tracking and actuator deflections are shown in Fig. 11 and Fig. 12.

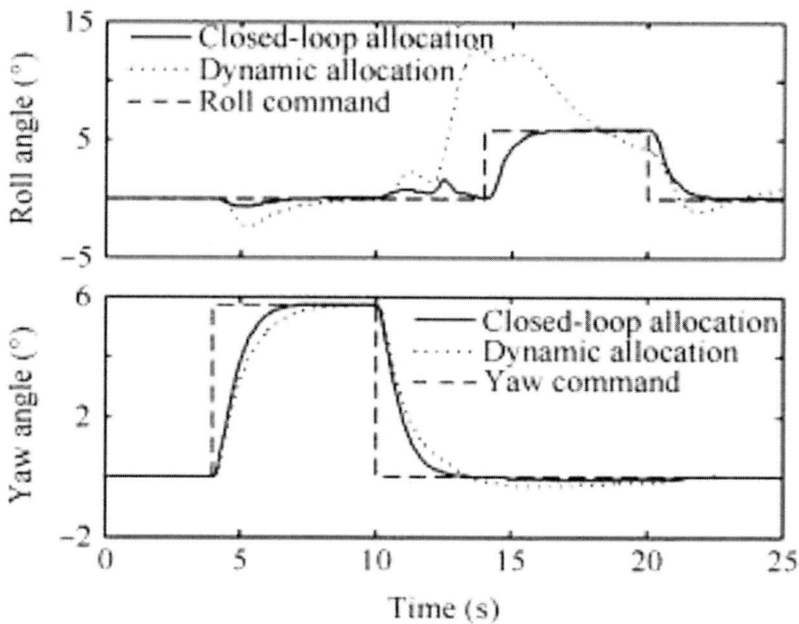

Figure. 11. Responses of lateral attitude with left aileron stuck.

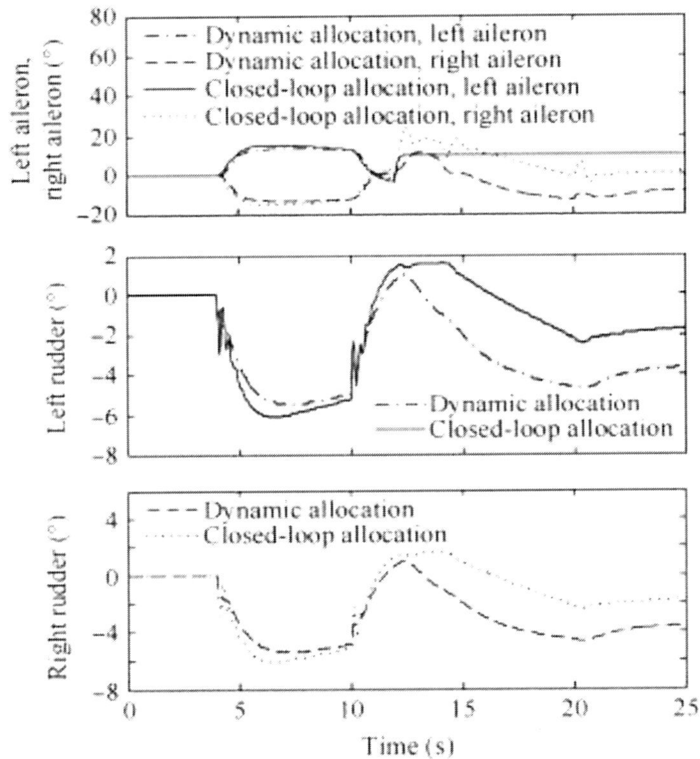

Figure. 12. Deflections of lateral actuators with left aileron stuck.

Fig. 11 shows that the desired lateral attitude commands, especially the roll command, cannot be tracked by the dynamic control allocation method after the failure happens. InFig. 12, we can see the right aileron deflects to counteract the disturbance from the locked left aileron and derive the desired roll moment using the proposed method. Meanwhile, the left and right rudders deflect to derive the desired yaw moment and counteract the disturbance from the roll channel. And the eigenvalues of $\boldsymbol{Vu}(k\ )$ are $\lambda_1^u = 1$ and $\lambda_2^u = 0.5$, and $\boldsymbol{u}(\infty) = [10° \ 10° \ -0.005° \ -0.005°]^T$. According to Corollary 2, the closed-loop dynamic control allocation system is stable. The actuator deflections in Fig. 12 are very close to the $\boldsymbol{u}(\infty)$ when the simulation time extends to 80 s.

Fig. 13 and Fig. 14 show the open-loop frequency responses of closed-loop dynamic control allocation system in two cases in both longitudinal and lateral directions. We can see the proposed method has enough gain and phase margins before losing stability.

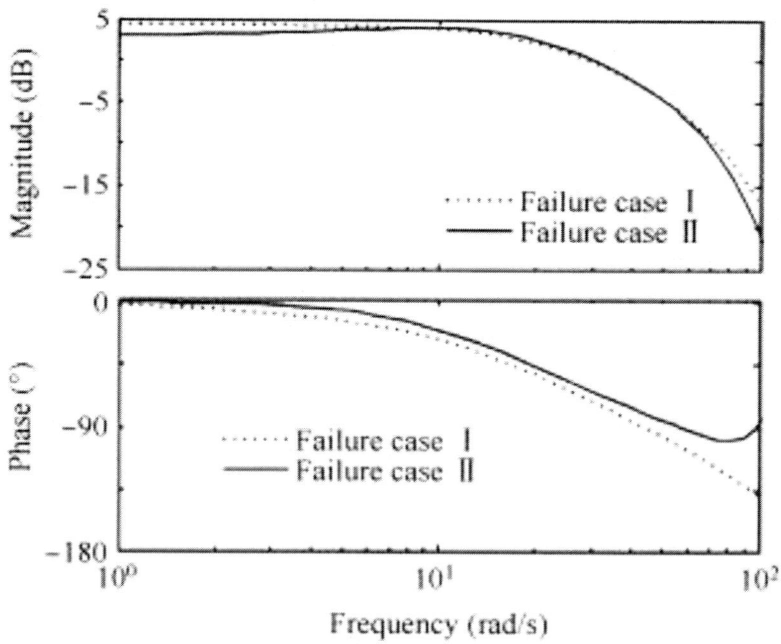

Figure. 13. Frequency responses of closed-loop dynamic control allocation in two failure cases in longitudinal direction.

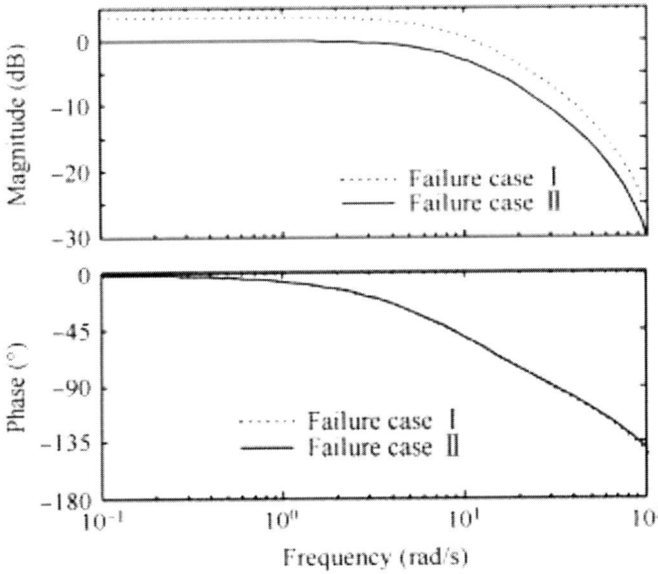

Figure. 14. Frequency responses of closed-loop dynamic control allocation in two failure cases in lateral direction.

5. CONCLUSIONS

In this paper, a closed-loop control allocation approach is proposed for aircraft with multiple actuators. By combining closed-loop configuration with dynamic control allocation, the proposed method guarantees that the closed-loop system is stable in the absence or presence actuator failures, and the actuators work in the respective frequency domain. The CRW aircraft model example demonstrates the effectiveness of the proposed approach.

The actuator dynamics presented here does not contain actuator saturation. Therefore, the closed-loop dynamic control allocation approach with actuator saturation is our future research.

ACKNOWLEDGEMENT

This study was supported by Program for New Century Excellent Talents in University (NCET-10-0032).

REFERENCES

1 Snell SA, Enns DF, Garrard WL. Nonlinear inversion flight control for a supermaneuverable aircraft. J Guid Control Dyn 1992;15(4):976–86.

2 Enomoto K, Yamasaki T, Takano H, Baba Y. Guidance and control system design for chase UAV; 2008. Report No.: AIAA- 2008-6842.

3 Ezal K, Pan Z, Kokotobic´ PV. Locally optimal and robust backstepping design. IEEE Trans Autom Control 2000; 45(2):260–70.

4 Sonneveldt L, Chu QP, Mulder JA. Nonlinear flight control design using adaptive backstepping. J Guid Control Dyn 2007;30(2):322–36.

5 Bodson M. Evaluation of optimization methods for control allocation. J Guid Control Dyn 2002;25(4):703–11.

6 Shtessel YB, Buffingtion JM, Banda SS. Tailless aircraft flight control using multiple time scale reconfigurable sliding modes. IEEE Trans Control Syst Technol 2002;10(2):288–96.

7 Durham WC. Constrained control allocation. J Guid Control Dyn 1993;16(4):717–25.

8 Tang SY, Zhang SJ, Zhang YL. A modified direct allocation algorithm with application to redundant actuators. Chin J Aeronaut 2011;24(3):299–308.

9 Buffingtion JM, Enns DF. Lyapunov stability analysis of daisy chain control allocation. J Guid Control Dyn 1996;19(6):1226–30.

10 Petersen JAM, Bodson M. Constrained quadratic programming techniques for control allocation. IEEE Trans Control Syst Technol 2006;14(1):91–8.

11 Poonamallee VL, Yurkovich S, Serrani A, Doman DB. A nonlinear programming approach for control allocation. In: Proceedings of the 2004 American control conference; 2004. p. 1689–94.

12 Ha¨rkega˚rd O. Dynamic control allocation using constrained quadratic programming. J Guid Control Dyn 2004;27(6):1028–34.

13 Arun Kishore WC, Sen S, Ray G, Ghoshal TK. Dynamic control allocation for tracking time-varying control demand. J Guid Control Dyn 2008;31(4):1150–7.

14 Yang LY, Fan YM, Shao S, Zhong YW, Shen GZ. Flight control based on the moment compensation and control allocation for aircraft with multiple control effectors. Control Theory Appl 2010;27(5):631–5 [Chinese].

15 Stevens BL, Lewis FL. Aircraft control and simulation. New York: John Wiley & Sons; 2003.

16 Durham WC, Bordingnon KA. Multiple control effector rate limiting. J Guid Control Dyn 1996;19(1):30–7.

17 Gai WD, Wang HL, Li DW. Flight dynamic modeling and analysis for the canard rotor/wing UAV. Acta Aerodyn Sin 2012;30(2):244–9 [Chinese].

18 Gai WD, Wang HL, Guo TF, Li DW. Modeling and LPV flight control of the canard rotor/wing unmanned aerial vehicles. In: Proceedings of second international conference on artificial intelligence, management science and electronic commerce; 2011. p. 2187– 91.

19 Bass SM, Thompson TL, Rutherford JW, Swanson S. Low-speed wind tunnel test results of the canard rotor/wing concept; 1993. Report No.: AIAA-1993-3412.

20 Mitchell CA, Vogel BJ. The canard rotor wing (CRW) aircraft: a new way to fly; 2003. Report No.: AIAA-2003-2517.

CHAPTER 5

STEPS TOWARDS SCALABLE AND MODULARIZED FLIGHT SOFTWARE FOR UNMANNED AIRCRAFT SYSTEMS

Johann C. Dauer[1*], Lukas Goormann[1] and Christoph Torens[1]

[1] German Aerospace Center (DLR), Institute of Flight Systems, Braunschweig, Germany

ABSTRACT

Unmanned aircraft (UA) applications impose a variety of computing tasks on the on-board computer system. From a research perspective, it is often more convenient to evaluate algorithms on bigger aircraft as they are capable of lifting heavier loads and thus more powerful computational units. On the other hand, smaller systems are often less expensive and operation is less restricted in many countries. This paper thus presents a conceptual design for flight software that can be evaluated on the UA of convenient size. The integration effort required to transfer the algorithm to different sized UA is significantly reduced. This scalability is achieved by using exchangeable payload modules and a flexible process distribution on different processing units. The presented approach is discussed using the example of the flight software of a 1.5 kg unmanned helicopter and an equivalent of 14 kg. The proof of concept is shown by means of flight performance in a hardware-in-the-loop simulation.

KEYWORDS

UAV, UAS, Flight Control Software, Unmanned Helicopter

1. INTRODUCTION

Typically, unmanned aircraft (UA) that are designed to fly in urban settings have to cope with computationally costly algorithms. The fusion of all sensors for state estimation, flight control and mission management is already complex although these are only the basic modules. Furthermore, obstacle sensing has to be included and environmental mapping integrated, which creates a model of the environment and its impact on the mission. On-board path planning is necessary if the area of interest could cause a deviation from the predefined flight path.

These UA are supposed to operate automatically even with a temporary loss of the data link to the ground control station (GCS). Thus, all these algorithms have to run on-board. This conglomerate of on-board software components, which include, among others, the flight controller, mission management, sensor fusion, path planning and the required middle ware, is what we refer to as *flight software*.

Even for UA at the scale of dozens of kilograms which can carry multicore processors and gigabytes of RAM this is still an integration challenge. For small UA which only weigh around a few kilograms this task increases in difficulty. On smaller platforms, a mission specific rigging tackles the lack of lifting capability. This rearrangement of the avionics, however, requires a flexible and scalable software architecture.

Consider, for example, the development of certain algorithms of obstacle detection or on-board path planning, as presented in [1]. These algorithms are computationally costly and thus evaluation is more convenient on bigger systems as the code optimization is less extensive.

However, these algorithms gain in importance the smaller the system becomes, as these systems fly at lower altitudes and are thus more likely to actually come across obstacles. Furthermore, smaller systems often underlay less restrictive governmental regulations and can be operated with less organizational effort. Additionally, the components of the avionics and the test-bed itself are less expensive. Thus, both approaches have advantages for research.

This paper presents a software concept where algorithm development can be done on a UA of a convenient size. The fully integrated algorithm can be directly used on UA of different size.

Globally, this scaling is limited by two aspects. First, the guidance, navigation and control algorithms can only be applied to vehicles of a scale where the algorithm's properties do not change fundamentally. Consider, for example, a flight controller as presented in [2]. Here it has been demonstrated that an adaptive control algorithm can be applied to different aircraft, provided an identical interface definition and a vehicle dependent control allocation are used. Second, the availability of hardware imposes restrictions due to limits in downsized equipment including CPU or sensors.

Figure 1. Mid-size helicopter, MTOW 14kg

For the implementation our proposed concept, the interactions between different CPU architectures have to be considered. As these aspects vary significantly depending on the architecture chosen, they are not discussed in detail here and are left for a careful analysis during the implementation stage.

It is desirable for the smaller system to approach the overall computational power of the bigger system as closely as possible. Depending on the electronic components available, this might only be possible if multiple miniature processing units of reduced power are introduced. The overall trade-off is therefore to balance the number of units and the complexity implied by multiple parallel processing units. As a result, the software has to be distributable to different computational units optionally.

The property of the presented software architecture which enables its direct use on UA of different sizes, the exchange of hardware units (processing unit and periphery) and an optional distribution in different processes is referred to as *scalability* in this paper. As will be seen later, key elements are the communication, time-synchronization modules and a "hardware equivalent" implementation of the avionics.

The concept is illustrated by refactoring the flight software of the 14 kg helicopter ARTIS (Autonomous Rotorcraft Test-bed for Intelligent Systems) of the German Aerospace Center shown inFigure 1. The software is transferred to a small helicopter based on a T-Rex 450 platform with an MTOW (maximum take-off weight) of around 1.5 kg, see Figure 2. The description in this paper is limited to the software environment. Changes needed to the guidance, navigation and control algorithms themselves are not outlined here as the concept is designed so as to be independent from the specific algorithm implementation.

Figure 2. Small-size helicopter, MTOW 1.5kg

The next chapters are organized as follows: After a brief overview of the related work, the general concept of the flight software will be discussed. Afterwards, an avionics concept is outlined for both helicopters. In the third chapter the software implementation based on the presented hardware design is briefly outlined. The implementation is tested and evaluated in the final chapter using a hardware-in-the-loop simulation which includes the avionic systems in the loop.

2. RELATED WORK

Flight software for UA has been the focus of research for the past fifteen years. In many publications it has been shown that it has similarities to both robotic and space applications. A well-structured overview of related aspects of robotic and space software is given in [3]. The following section shall thus be confined to aspects of architectural representation which affect the scaling problem of this paper.

Many papers present architectures as abstractions in layers. One of the basic architectures is a three layer abstraction [4] which became the basis of many proceeding architectures. The layer called reactive controller forms the basis which is overlaid by the plan execution which coordinates the mission in a reactive sense. The deliberative layer includes the third and less frequent task imposing most of the computational costs.

Research has been published which specifies these layers and adapts them according to several requirements. Incorporating decision making as an explicit layer results in a four layered architecture [5]. Within UA applications there is often the need for a parallel process, which deals with environmental awareness. Koo et al. presented a multilayer description consisting of a strategic planner, tactical planner and a controller supervised by a vision layer in [6]. A conceptual extension with respect to requirements of sensor data and which considers risk management is presented in [3] and its application in collision avoidance is proposed. Alternatively to segmentation into architectural layers, these modules can also be categorized into cognition layers (Perception, Cognition, and Action) from a situational awareness perspective [7].

Further focus has been placed on the communication between the modules as presented in [8], for example, where inter process communication is modelled as peer-to-peer pipes. Every process has one message queue. In general, the modules responsible for this data exchange are often included within the middleware. These aspects have previously been addressed using middleware like CORBA (Common Object Request Broker Architecture) in [9]. In [10], also based on CORBA, the authors focused on an avionics system divided into two parts – flight control and vision payload.

Due to its platform, independent *Interface Definition Language,* CORBA has advantages regarding communication within systems of a hybrid nature in relation to platform and programming language. This advantage comes at the cost of an increased implementation effort and a more complex system due to the translation workload in the multi-language environment. In [7], ROS (Robot Operating System) is chosen as middleware. However, ROS does not support clock synchronization and imposes its own architecture on all software modules. It is thus not suitable for introduction of scaling in existing systems.

In [11] and [12], a hub-like communication management is presented for the Berkely Plane UA. Here, communication includes a state broadcast between multiple UA. López et al. build a specific middleware for UA avionics [13]. In their approach, the middleware executes and manages the other modules. This approach has the drawback that the middleware becomes mandatory for each module.

Based on that middleware, a service-oriented architecture is described in [14] and [15].

Aspects of the software architecture in the sense of intelligence abstraction shall not be discussed here, as there is extensive literature as shown above. The modules of the software presented in this paper can easily map to some of the previously mentioned architectures.

A focus of this paper is the scalability in light up to middle weight applications. To the knowledge of the authors, there is no complete architecture representation for the scalability and module interconnection from a perspective for integration in UA of different scales.

3. SOFTWARE CONCEPT

The overall aim of an UA, such as the one presented in the introduction, is automated flight. It flies through areas with obstacles that are not necessarily known exactly a priori and automatically accomplishes certain missions within that area. Additionally, the vehicle is equipped with mission-dependent payload.

The following gives a short overview of the smallest set of mandatory higher level modules that have to be integrated into the flight software. The four components for the flight software are sensor fusion, flight controller, trajectory generation and mission management.

The sensor fusion gathers sensor information and estimates the current state of the UA using estimation algorithms like Kalman filters. A trajectory generator calculates the time-wise evolution of the position and attitude command of the UA, cf. [16]. The flight controller is an implementation of the control algorithms which can be arranged in a control cascade, consisting of rate, attitude, velocity and path control; cf. [17] or equivalent. Finally, the mission management is responsible for the coordination of mission elements. It receives the mission from the GCS and generates commands for the flight control system [18].

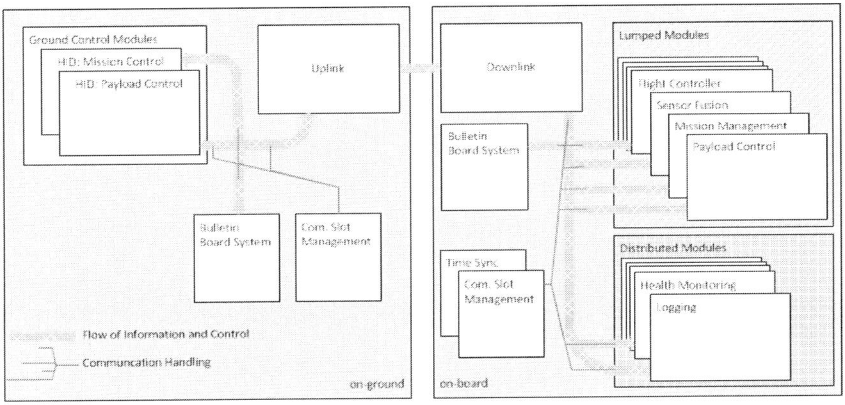

Figure 3. Information flow of the flight software and ground control station

These components are dependent on the current equipment and mission of the vehicle. The sensor fusion incorporates all primary sensors, which might, in general, not be exchanged regularly. However, if applicable, a particular payload can be included as well. Vision can support navigation and be included in the sensor fusion, as presented in [7]. The controller might be dependent on a certain payload as well, for example, with swing loads.

Additionally, modules for logging and health monitoring are required. These modules ensure the supervision of the safety of the.

All these components are introduced in Figure 3. The figure shows the two aspects of what is often referred to as an "unmanned aircraft system", the (ground) control station (GCS) and the UA itself. Both sides can be distributed on different computers or computational units. The GCS can be separated into UA operations and the operation of the payload.

Especially for small scale UA, the on-board components are distributed on different computational units, as will be shown later. The figure hence categorizes modules as being lumped or distributed. The lumped modules run on one computational unit alone, while the distributed work on different units, optionally requiring an information exchange.

The communication flow shown in the figure illustrates the concept of inter-module communication. There is one uplink between the ground control station and the UA. The reason is that on-board hardware interfaces and weight resources are limited on small systems. However, for bigger systems additional uplinks may be equipped implying a dedicated management of suitable up-and downlink channel selection. In that case, payload control might directly communicate with payload components on-board. For scalable systems, these links are optional and therefore are not shown.

There are actually two different aspects of the communication used for inter-module data exchange. A subscription system analogous to a bulletin board and direct data links. The communication might be implemented using a suitable middleware as highlighted in the related work section. However, in this case, for three main reasons, the communication was implemented from scratch. This was done in order to fulfil the requirements for this particular application, gain knowledge and, most importantly, make the communication module as light as possible. As communication and time synchronization are the major aspects of the scalability they are described in more detail below.

Generally, distribution across different computational units can be carried out at any of the communication pipes. However, some criteria have to be considered. Not fulfilling these criteria also corresponds to the identified hazards of a functional hazard analysis (FHA) [26] that has to be performed for certification considerations (Section 0). The criteria consist of the following items:

- Communication will cause additional delays to the data. Software modules may only be separated hardware wise if the connection ensures that the client module receives the required data in time. In the example shown below, the flight controller receives the state data of the UA delayed by an additional network channel. This requires the controller to have a sufficient time-delay margin.
- The amount of data to be transferred is smaller than the bandwidth of the hardware interconnection.
- The overhead introduced by the additional layers that the data runs through does not cause too extensive a computational burden.

On the other hand a separation has to be carried out:

- if the subset of modules would otherwise exceed the capacity of one computational unit,
- if the amount of interfaces required for peripheral components exceeds the number of those available.

Furthermore, separation can enhance the security of the system because the core system can be separated from a rather experimental payload. More sophisticated but experimental algorithms might be separated from more robust ones as a fall back solution, as was done for the sensor fusion described in [19].

Logging shall not be discussed in more detail, as there are numerous ways of completing the task, which are strongly dependent on the platform in use. It has to be kept in mind that logging can, however, place significant loads on the system if the gathered data is extensive. If debugging of a particular algorithm requires costly logging, this might actually be one criterion for its development on a bigger system with sufficient computational power to handle the logging.

The remaining support modules needed for the presented architecture are outlined in the following sections.

3.1. Scalability Aspect: Health Monitoring

Health monitoring is one support module of the flight software. Figure 4 shows its concept. Callbacks can be registered which define health data, states and corresponding actions: A data callback is registered which the health monitoring uses to gather the required data at a specified frequency. A corresponding criterion is registered which is used to determine the state of the module based on the data just gathered. Finally, a number of pairs is registered which consist of the possible states and callback to their corresponding actions.

Using this approach, the monitoring remains independent of the hardware, each piece of equipment and desired module can register itself and will automatically be included within the monitoring.

The task of the health guard is to check the existence of the health monitoring and in case of a crash to restart it. It must therefore be informed of any registering process to be able to recreate the health monitoring in its complete state before a crash. Vice versa, the health monitoring checks the existence of the health guard. The health monitoring module offers a data structure for logging and transferring to the GCS.

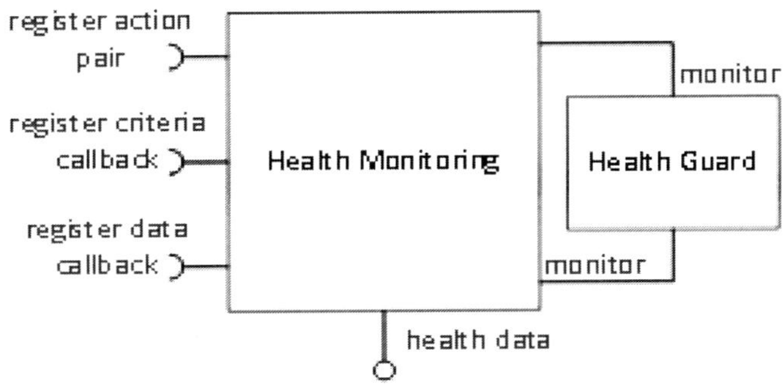

Figure 4. Health monitoring including its guard (UML semantic)

3.2. Scalability Aspect: Communication

The communication connects all the different modules and processes on a UA. We decided to build a communication module that manages the communication between the UA and its ground control station, between applications on distributed hardware on-board and with other UA. The communication module is designed so that it can easily be integrated into existing modules and interfaces.

The communication module has to fulfil several requirements. First of all it should be as fast and involve as few computational overhead as possible, as avoiding delays is crucial for many algorithms. On the other hand it needs to be secure and reliable. As it also separates different modules from each other, it may also be regarded as a security layer in some cases.

A third requirement is that the communication structure is sufficiently flexible. It shall be possible that communication partners change over time. New processes may be started or obsolete ones closed. None of these dynamic changes in the communication structure shall effect any other existing communication connection. The communication itself can be divided into four different types:

1. *Bulletin board data:* Periodically updated data which might be used as an input to several modules: Usually sent in small to mid-size packages, less than one kilobyte in size. New values override old values. If one update does not reach the receiver, sending the next update without recovering the missed transfer will typically not cause a failure. This data is published on a virtual bulletin board and can be subscribed to by all interested applications in the network.

2. *Single-shot messages:* Requests, commands or answers to requests, sent to a dedicated module: These are usually small packages, but are sent only once. Receiving these messages is crucial.

3. *Streamed data:* Updates for larger data which might be used as an input to several modules: In order to save bandwidth only incremental updates are sent with each step. It is critical that all messages reach the receiver in the correct sequence to avoid inconsistent states.

4. *File transfers:* Data that is too large to send in one package and need to be separated into several packages: Usually the receiving algorithms are not designed to interpret the data before they are completely and correctly received.

These four types are supported by the communication module. The implementation is based on so called topics. A topic is specified by an identifier as shown in Figure 5, which consists of three elements:

One identifier for the receiver, one identifier for the sender and one for the transmitted data type and semantic.

The identifiers for the sender and receiver are composed of an identifier for the hardware the application is running on and an identifier for the instance of the application unique on that hardware. The application identifier is generated dynamically at the start of the process and not predefined, e.g., during compile time. Thus, it is possible to execute multiple instances of the same application with unique identifiers on the same hardware.

So called slots are used to access the topics as sender or receiver. Each application can assign slots to existing topics. Typically, slots with the same topic identifier should always – after some delay of communication – hold the same data. Slots are divided into sending and receiving slots. For each topic there should usually be just one sending slot at the same time.

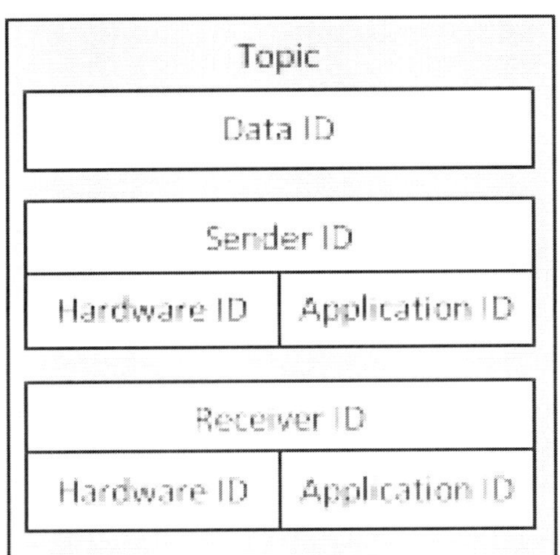

Figure 5. Topic Identifier

The sending slot determines if the connection needs to be secure. This is the case for single-shot messages, streamed data and file transfers. In that case the sending process is asynchronous and the calling function can be notified if any transmission error occurs. Furthermore, that slot can also distribute large data into several packages guaranteeing that the transmission will be done in sequence and completely.

Receiving slots will hold an actual data package until a new one is received. For the communication of streamed data and commands, receiving slots can also trigger events in order to notify connected algorithms of incoming new data. It is guaranteed that all notified algorithms may react to the event before the next data is received. For bulletin board data, the receiver identifier of one topic is set to "any interested entity" instead of a specific receiver. The data is then distributed to any receiving slot having the same topic ID. Behind the scenes this is ensured by subscription the alternative, a broadcast, would waste bandwidth.

A register of all known potential communication partners and their application names is held in each process. One process per hardware holds the so called "master" application register. The master register exchanges his knowledge of the actual network with all the other master registers on different processing units. It shares all news with the local registers.

Figure 6 shows an exemplary situation. The knowledge of the complete network is distributed, so the application register of Hardware Unit 1 and 2 will also be informed of the existence of the application at Unit 4 although there is no direct link. If the process holding the master register is closed or aborted – such as Application 1 of Unit 2 – any one of the other processes on the same hardware will automatically take its role. However, all communication will take place between reachable partners directly so that a failure in any other module or master register will not interrupt the communication.

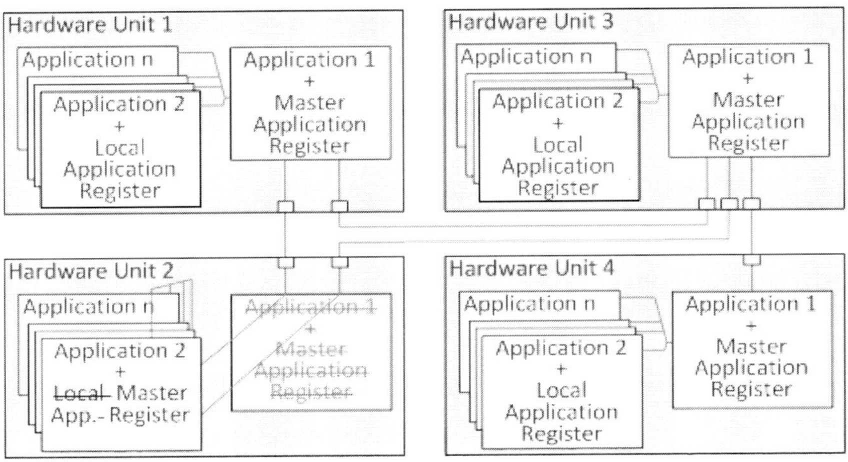

Figure 6. Communication Handling: Local and Master Application Register in different applications on distributed hardware of one network

The communication module also abstracts the technical implementation of the data links between the different hardware. It can send the data packages over (wireless) LAN as well as serial connection between two processing units.

The communication module also collects information about the communication taking place, such as the frequency and bandwidth between communication partners. That information can be used for routing as well as for the health monitoring module.

3.3. Scalability Aspect: Time Synchronization

The decisions and actions of an UA are usually based on the mission and the measurements of various sensors. To fuse the datasets of different sources correctly and to interpret them, it is essential to have consistent timestamps. Furthermore, concurrent information and commands can only be interpreted correctly if their timestamps can be correlated to the local on-board clock.

There are various ways to achieve consistent timestamps are mentioned in literature. If the delays in the data acquisition are known or negligible and processing is performed on the same computer, no further efforts are required. Thus, some UAS collect the sensor data on one processing unit connected to one clock before further processing [12]. For the presented scalable flight software, it is not necessarily known a priori how many computational units are used. Furthermore, we explicitly want to make time delay investigation possible where needed.

Some UAS use network servers and clients to synchronize the system clock of all the connected hardware either to each other or to GPS time [7]. However, this method can have some drawbacks. First, GPS might not be available as a reference all the time. Second, adjusting the clock might invalidate old data, time stamped before a time shift. Third, if the data exchanging partners are not all connected to each other constantly, as in cooperative UA scenarios, it might be impossible to synchronize all the UA correctly. For example, if two non-synchronized UA detect inconsistent phenomena at different times and report to a third, it might be impossible to determine the most recent.

If the time synchronization is done by a dedicated server, as for the precision time protocol (PTP) concept [20], a special handling should be used in cases where the server is not available, e.g., due to software or hardware failures.

Römer suggests estimating one function to convert the timestamps of different processing units for each pair of communication partners instead of adjusting the on-board clocks [21]. The estimation of the converting function is done in two steps.

First, the communication delay is estimated. This is achieved by sending the time difference between the local time of a package and the sending time in the remote package back to the sender. This is also done in the PTP, the network time protocol and various other time protocols [18, 20].

Second, timestamps of the remote system are collected, the communication delay is eliminated and the conversion function is estimated, for example, by linear regression. Some methods for the estimation are presented in [23]. The conversion functions are stored in each process and sorted by the related hardware unit of the communication partners, as shown in Figure 7. The first application on Hardware Unit 3 is to communicate with the application on Hardware Unit 1 and, therefore, to estimate the convert function for that hardware timestamps. The application on Hardware Unit 1 also communicates with an application on Unit 2 and, therefore, needs to estimate two convert functions, for Hardware Unit 2 and 3. Exchanged estimated functions between the communication partners can be used to determine initial values for the synchronisation and plausibility checks.

The package headers of the communication protocol already include a sending timestamp. Additionally, the estimation of communication delays is also important for the communication module, e.g., for route-selection. Thus, the time synchronization can easily be integrated into the communication module. Only an expansion of the package header with two 32 bit values for the last measured time difference between the sending time on the remote hardware and the last local receiving time is needed. If the communication traffic between two processing units is frequent enough, no further packages for time synchronization need to be sent.

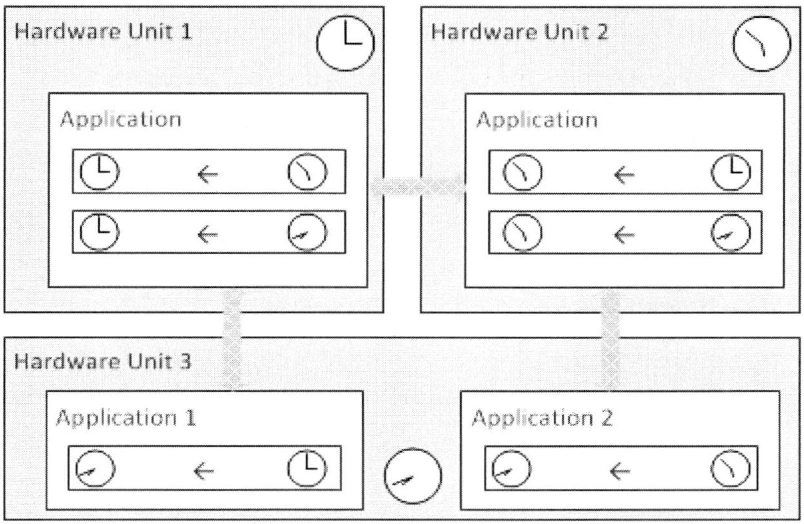

Figure 7. Time synchronized applications on different hardware

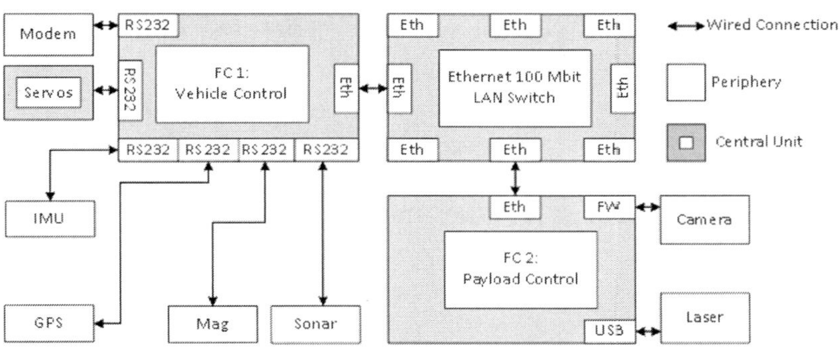

Figure 8. Interconnection block diagram of the mid-scale helicopter based on two PC104 systems

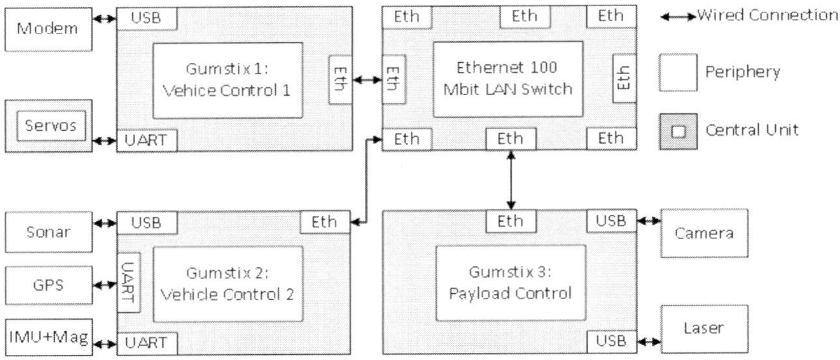

Figure 9. Interconnection block diagram of the small-scale helicopter avionics based on Gumstix boards

4. CONCEPT OF THE AVIONICS

The previous chapter showed the software concept. This section outlines the avionic systems for both helicopters. Afterwards, the implementation will outline the specific aspects for these avionics.

The avionics system of the 14 kg helicopter is based on an Ethernet interconnected PC104 system which is separated into one Linux-based payload computer (FC 2) and a QNX-based flight computer (FC 1), see Figure 8. Optional wireless LAN connections to the GCS are not shown.

The hardware concept of the miniaturized avionics is shown in Figure 9. In order for it to be compatible with the PC104 system, the hardware interconnection is established using a miniaturized 100Mbit Ethernet LAN switch. Three Gumstix form the computational units. Gumsitx are small processor boards equipped with ARM Cortex A8-based processors designed in an open source hardware concept.

Gumstix 1 handles the communication with the ground control station over a serial modem and is also connected to the safety electronics which on the one hand handles the actuator driving and on the other hand does the switching between automatic and manual mode for the safety pilot in case of failure. From an algorithmic perspective, Gumstix 1 handles all the loops of the flight control system and the mission management.

Gumstix 2 calculates the state estimation based on the sensors connected to this board. A Sonar is used to detect the height above ground during landing and take-off. It also uses a single or differential GPS. An integrated inertial measurement unit (IMU) and magnetometer (Mag) solution completes the basic sensor equipment. These first two processor boards are operated using QNX. The third Gumstix handles vision applications based on a Linux derivate. It can either be equipped with a small scale camera or a laser scanner with a weight of less than 350g.

This avionic system is mainly designed so as to be easily extendible with further LAN capable hardware modules as well as ease in exchange of different components. As stated in [24], flight software for space applications is tightly coupled due to limited recourses. This is true for small scale aircraft too. However, by designing the avionics as described here, flexibility in the sense of exchangeable modules is created. Whole payload modules can be added and exchanged, including the dedicated computing unit and not only the sensor itself. Only the compatibility of the software interfaces has to be ensured. As both avionic systems are based on the same concept, they are what we call *hardware equivalent*.

Note that this particular assignment of tasks to processing units is neither the optimal workload distribution nor optimal from the perspective of the algorithm requirements. Separating flight control system and sensor fusion, for example, will cause additional delays to the helicopter state data due to the helicopter state being transmitted over LAN. The number of hardware interfaces, however, demands a separation so that all the sensors can be equipped to one board and thus the servo driving board to another.

5. IMPLEMENTATION

In this chapter an implementation of the software concept is outlined that eases the exchange of the modules between different scales of avionics. The software runs in separate processes on the PC104 avionics system and is distributed to a set of processing units on the Gumstix variant. A further subdivision into even more processes is certainly possible. Nevertheless, this refactoring process is meant to be continuous. Separation into sub-processes will only occur if certain hardware requirements demand it in order to maintain a complexity as low as possible.

Figure 10 presents the structure of the software. The colouring is identical to Figure 3. Every process contains instances of health monitoring and logging. The interfaces symbolize the exchanged data transmitted either over LAN or in shared memory between processes. The flight control and mission management process is also responsible for ground control communication. In this role, it also coordinates the remaining processes using the control and observer modules.

The control and observer modules make sure that every command received from the ground station is correctly transferred to the remaining processes (ctrl) and checks for the correct reaction on the other process side. Health data (hlth) is transferred to the management process as well as sensor fusion data (sfd) and vision or payload data (vd).

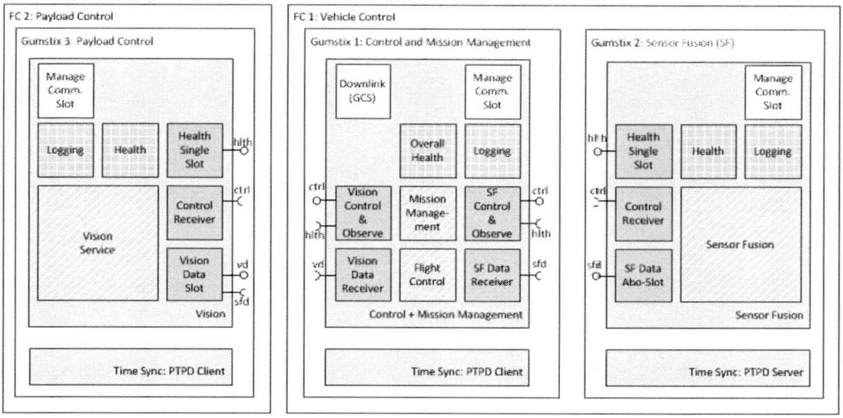

Figure 10. Implementation of the flight software for scaling between three processing units in the case of the Gumstix avionics and two for the PC104 variant

The time synchronization of the different processor units is implemented using a daemon-based implementation of the PTP according to IEEE 1588 [20]. This implementation does not enable time synchronization with the ground station via the uplink and other cooperating UA. This will be enabled as soon as time synchronization is fully included within the communication framework as previously presented in the time synchronization chapter. The achievable synchronization accuracy is below 1 ms1 ms according to the self-diagnostics of the synchronization processes. Health monitoring of the communication and estimation of the transfer delay is hence possible.

6. PROOF OF CONCEPT

This chapter discusses results of a hardware-in-the-loop (HIL) simulation as an example of the tests performed. Figure 11 shows the general test setup of the HIL setup. The complete avionic computational units are integrated as they are in flight. The real time simulator measures the servo signals of the helicopter, calculates the corresponding flight mechanical responses and generates sensor emulated signals based on these computations. The protocols of the sensors are completely emulated and are connected via the same hardware interfaces as the real sensors would be.

From the computational units and the software perspective, there is thus no difference compared to real flight. To achieve identical flight behaviour, the flight mechanical model of the 14kg midiARTIS helicopter is used for both avionics. The flight mechanics consist of a linear hover model with eight degrees of freedom. Nonlinear extensions cover external forces. Furthermore, flight test-based error models of the sensors as well as identified actuator models are simulated.

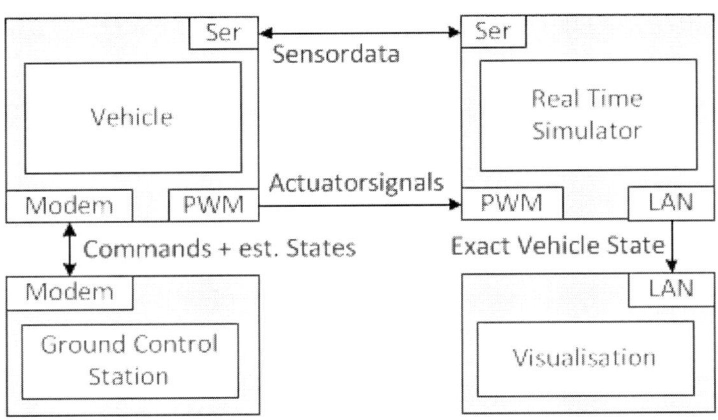

Figure 11. Hardware-in-the-loop-simulation

This setup enables identical flights for the different avionic and software architectures and real comparison of the variants' performance is possible. The vehicle is operated using the ground control station, again identical to a real flight scenario.

For the sake of brevity, we limit ourselves to one example path shown in Figure 12. The reference implementation, denoted by PC104 single, is the original implementation. It contains guidance navigation and control algorithms identical with the scalable version. It differs from Figure 10 by removing communication classes in a single process implementation and missing time synchronization components. The PC104 distributed version consists of the scalable software architecture, where the sensor fusion and flight control processes run on the flight computer of the PC104 avionics. It is apparent that there is no significant difference in flight performance compared to the original implementation.

The final implementation is the multi-process version, compiled for the ARM Gumstix avionics presented in Figure 9. The flown path is nearly the same as the original. The slight difference from the reference implementation is caused by the additional delay due to the LAN interconnection between the sensor fusion and flight control system. The delay was measured to remain between 2 and 8 ms.

6.1. Certification Considerations

ARTIS is an experimental vehicle, therefore, the certification of its flight software using correspondent standards [25, 26] is not the main focus here. Nonetheless, a preliminary FHA [25] adds three main hazards to the system, compared to the previously existing monolithic system architecture. The hazards have been outlined in Chapter 3 as required criteria: (1) too large a communication delay is introduced, (2) communication data exceed available bandwidth, (3) too large a communication overhead is introduced. Additionally, the scalable architecture alters the overall system design, but the software level classification [26] of previously existing modules is not affected.

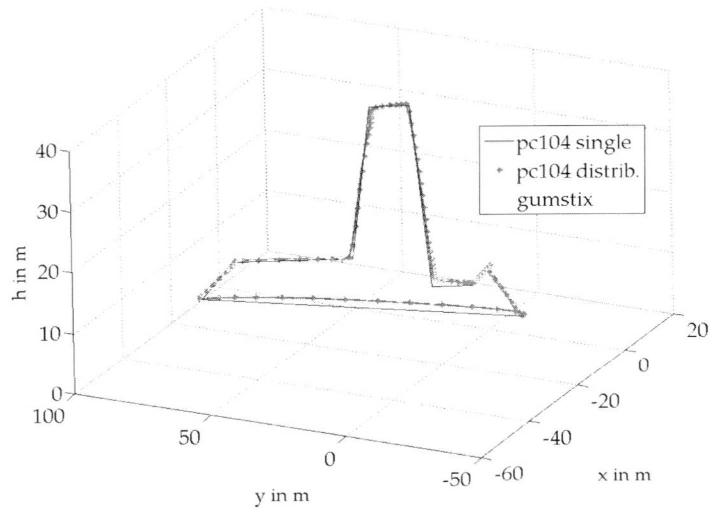

Figure 12. Example path flown with different avionics and software combinations

With these FHA results in mind, the communication module, as well as aspects of time synchronization, fall into the DO-178C software level A [26], where a failure might cause multiple fatalities with a loss of the aircraft. Since safety considerations are not the core of this papers' contribution, only a qualitative analysis will be provided for the proof of concept. Further reading on recent certification considerations of ARTIS can be found in [27].

Tests and analyses as well as HIL simulations showed the communication delay to remain below 8 ms. Therefore, the resulting flight path only differs marginally. Measurements showed that the bandwidth used also stays well below the hardware specification for the network traffic. The overhead for managing the network traffic has hence no impact on the flight performance.

The health module has no negative impact on the certification aspects of the architecture used, since it imposes no active action on other modules. However, future work will elaborate on the beneficial aspects of such a module.

7. CONCLUSION AND FUTURE WORK

This paper presents a software concept for the flight of unmanned aircraft. A focus of this architecture is an application to vehicles of significantly different sizes. The software is thus focused on scalability and the optional distribution to different numbers of computers. Health monitoring, communication framework and time synchronization were outlined in depth.

It has been shown in a hardware-in-the-loop simulation that this software concept achieves equivalent flight performance compared to a monolithic implementation. It could thus be shown that the presented software concept is feasible for both scales of helicopter and will make the validation of future unmanned aircraft research significantly easier. Depending on the complexity and organizational demands of algorithms, the most convenient ones can be chosen. As soon as the algorithm has been integrated into the system, it will immediately work on the other variants as well, provided that the computational units can manage the algorithm's complexity and the algorithm handles the different UA flight properties.

Hardware concepts for two sizes of helicopter, a 14 kg and a 1.5 kg version, are presented. The avionic systems are based on x86 PC104 on the one hand and ARM Cortex A8 on the other. The scalability of the implementation of the presented software architecture is outlined in the example of both avionic systems.

It remains for future investigation to examine how an automatic priority setting of processes could relax computational demands on the small scale systems further. It is very possible that background and even distributed tasks can further increase the capacity of the small scale multiple computer system.

8. ACKNOWLEDGEMENTS

The authors gratefully acknowledge the help of the team members of the unmanned aircraft department, especially Jörg Dittrich, Sven Lorenz and Florian Adolf, who provided support and helpful discussions.

REFERENCES

1. Muratet L, Doncieux S, Briere Y, Meyer JA (2005) A contribution to vision based autonomous helicopter flight in urban environments. Robotics and Autonomous Systems, vol. 50, 195-229.

2. Mühlegg M, Dauer J, Dittrich J, Holzapfel F (2013) Adaptive Trajectory Controller for Generic Fixed-Wing Unmanned Aircraft. Advances in Aerospace Guidance, Navigation and Control, Springer Berlin Heidelberg.

3. Narayan PP, Wu PP, Campbell DA, Walker RA (March 20-21 2007) An intelligent control architecture for unmanned aerial vehicles in the national airspace system. 2nd Australasian Unmanned Air Vehicle Systems Conference. Melbourne, Australia.

4. Gat E (1998) Three Layer Architectures. In: Kortenkamp D, Bonasso R.P, Mruphy R editors. Artificial Intelligence and Mobile Robots. Cambridge: AAAI Press/MIT Press. pp. 195-210.

5. Boskovic JD, Prasanth R, Mehra RK (May 8-10 2002) A Multi-Layer control Architecture for Unmanned Aerial Vehicles. American Control Conference. Anchorage.

6. Koo TJ, Sinopoli B, Sangiovanni-Vincentelli A, Sastry S (Augsut 22-27 1999) A Formal Approach to Reactive System Design: Unmanned Aerial Vehicle Flight Management System Design Example. International Symposium on Computer Aided Control System Design. Hohala, Hawaii.

7. Tomic T, Schmid K, Lutz P, Dömel A, et al. (2012) An Extensible UAV Research Platform for Urban Search and Rescue Missions. IEEE robotics and automation magazine, special issue on aerial robotics and the quadrotor platform.

8. Reeves RE (October 10-12 2005) An Overview of the Mars Exploration Rovers Flight Software. IEEE International Conference on Systems, Man and Cybernetics, Hawaii, USA.

9. Doherty P, Haslum P, Heinzt F, Merz T, et al. (June 23-25 2004) A Distributed Architecture for Autonomous Unmanned Aerial Vehicle Experimentation. Proceedings of the 7th International Symposium on Distributed Autonomous Robotic Systems. Toulouse, France.

10. Paunicka J, Corman D, Mendel B (May 2-4 2001) A CORBA-based middleware solution for UAVs. Fourth IEEE International Symposium on Object-Oriented Real-Time Distributed Computing. Magdeburg, Germany.

11. Ryan A, Xiao X, Rathinam S, Tisdale J, et al. (August 21-24 2006) A Modular Software Infrastructure for Distributed Control of Collaborating UAVs. Proceedings of the AIAA Conference on Guidance, Navigation and Control, Ketstone, Colorado.

12. Tisdale J, Ryan A, Zennaro M (October 2-6 2006) The Software Architecture of the Berkeley UAV Platform. International Conference on Control Applications. Munich, Germany.

13. López J, Royo P, Pastor E, Barrado C, et al. (November 26-30 2007) A Middleware Architecture for Unmanned Aircraft Avionics. Proceedings of the 2007 ACM/IFIP/USENIX international conference on Middleware companion. Newport Beach, California: ACM.

14. Pastor E, Lopez J, Royo P (June 2007) UAV Payload and Mission Control Hardware/Software Architecture. IEEE Aerospace and Electronic Systems Magazine, Vol.22, No.6. pp. 3-8.

15. Pastor E, Barrado C, Royo P, Lopez J, et al. (2009 Jan) An open architecture for the integration of UAV civil applications. in T.M. Lam (Ed.), Aerial Vehicles. InTech. pp. 511–536.

16. Lorenz S, Adolf FM (2011) A Decoupled Approach for Trajectory Generation for an Unmanned Rotorcraft. In: Holzapfel F, Theil S (editors) Advances in Aerospace Guidance, Navigation and Control. Berlin: Springer-Verlag.

17. Lorenz S (2011) Open-Loop Reference Systems for Nonlinear Control Applied to Unmanned Helicopters. Journal of Guidance, Control, and Dynamics. Volume 35, No. 11, pp 259-269.

18. Adolf F, Thielecke F (2007) A Sequence Control System for Onboard Mission Management of an Unmanned Helicopter. AIAA Infotech Aerospace Conference. Sonamo, CA

19. Coopmans C, Han Y (2009) AggieAir: An Integrated and Effective Small Multi-UAV Command, Control and Data Collection Architecture. Proceedings of the 2009 ASME/IEEE International Design Engineering Technical Conferences & Computers and Information in Engineering Conference, San Diego, CA.

20. IEEE Std 1588-2008, IEEE Standard for a Precision Clock Synchronization Protocol for Networked Measurement and Control Systems,

21. Römer K (October 2001) Time synchronization in ad hoc networks. Proceedings of MobiHoc 2001, Long Beach, California.

22. Mills D L (October 1991). Internet time synchronization: The network time protocol. IEEE Transactions on Communications, 39(10):1482–1493.

23. Römer K, Blum P, Meier L (2005) Time synchronization and calibration in wireless sensor networks. Handbook of Sensor Networks: Algorithms and Architectures, I. Stojmenovic, Ed. John Wiley & Sons, Hoboken, NJ, pp. 199-237.

24. Dvorak D, Rasmussen R, Reeves G, Snacks A (March 25 2000) Software Architecture Themes in JPL's Mission Data System. Proceedings of IEEE Aerospace 2000 conference. Big Sky, Montana.

25. RTCA Std DO-178C/ED-12C (2011) Software Considerations in Airborne Systems and Equipment Certification.

26. SAE Std ARP4761 (1996) Guidelines and methods for conducting the safety assessment process on civil airborne systems and equipment.

27. Torens C, Adolf FM (2013) Software Verification Considerations for the ARTIS Unmanned Rotorcraft. 51st AIAA Aerospace Sciences Meeting. 1st Software Challenges in Aerospace Workshop, Grapevine, TX, USA. ISBN 978-1-62410-181-6.

CHAPTER 6

UAV FLIGHT CONTROL SYSTEM BASED ON AN INTELLIGENT BEL ALGORITHM

Huangzhong Pu[1], Ziyang Zhen[1*], Ju Jiang[1] and Daobo Wang[1]

College of Automation Engineering, Nanjing University of Aeronautics and Astronautics, Nanjing, Jiangsu, China

ABSTRACT

A novel intelligent control strategy based on a brain emotional learning (BEL) algorithm is investigated in the application of the attitude control of a small unmanned aerial vehicle (UAV) in this study. The BEL model imitates the emotional learning process in the amygdala-orbitofrontal (A-O) system of mammalian brains. Here it is used to develop the flight control system of the UAV. The control laws of elevator, aileron and rudder manipulators adopt the forms of traditional flight control laws, and three BEL models are used in above three control loops, to on-line regulate the control gains of each controller. Obviously, a BEL intelligent control system is self-learning and self-adaptive, which is important for UAVs when flight conditions change, while traditional flight control systems remain unchanged after design. In simulation, the UAV is on a flat flight and suddenly a wind disturbs it making it depart from the equilibrium state. In order to make the UAV recover to the original equilibrium state, the BEL intelligent control system is adopted. The simulation results illustrate that the BEL-based intelligent flight control system has characteristics of better adaptability and stronger robustness, when compared with the traditional flight control system.

KEYWORDS

Brain Emotional Learning, Unmanned Aerial Vehicle, Flight Control

1. INTRODUCTION

Unmanned aerial vehicles (UAVs) are widely used in military and civilian applications. In recent years, several nonlinear control methods, such as back-stepping, the Lyapunov function and nonlinear dynamic inversion, have been applied in small UAV flight control. Back-stepping techniques can derive air speed and roll control commands from known heading and air speed control laws which explicitly account for the heading rate and air speed constraints of the UAV. A miniature fixed-wing 6-DOF UAV model is used to show the effectiveness in trajectory tracking control [1].

Furthermore, three different approaches based on the state dependent Riccati equation, Sontag's formula and aggressive selection from a satisfying control set are proposed to design the heading and air speed control commands for this UAV [2]. The design of the formation control laws for YF-22 include inner and outer loop design for two aircraft in the formation - the outer loop scheme is based on feedback linearization, while the inner loop scheme is based on a root locus-based approach - and experimental results validate the performance [3]. The above methods improve the attitude and trajectory control performance of UAVs, however, the presented control methods are too complex to design and usually utilize a nonlinear model of a UAV.

Emotions have a strong faculty for decision making. In past decades, modelling of emotion has attracted the attention of many researchers both in cognitive psychology and the design of artificial systems. A network model which simulates the brain emotional learning (BEL) mechanism of mammalians is designed by Moren and Balkenius [4-5]. The BEL model is a computational model which mimics the amygdala, orbitofrontal cortex, thalamus, sensory input cortex and other parts of the brain.

From the BEL model being proposed, it was soon applied into control systems of real engineering fields, termed brain emotional learning based on an intelligent controller - originally proposed by Lucas [6]. In recent years, BEL controllers have proved to have good robustness and uncertainty handling properties when applied in many engineering systems, such as simo overhead travelling cranes [7, 8], switched reluctance motors [9], plant level systems [10], alarm systems [11], micro-heat exchangers [12], flight simulation servo systems [13, 14] and other uncertain nonlinear systems [15].

The BEL intelligent control scheme has proved to be effective and robust in real systems from the above references. Therefore, in this paper, the BEL intelligent control is initially used to improve the attitude control performance of a small unmanned aerial vehicle (UAV).

The organization of this paper is as follows. The dynamic model of the UAV and the biological structure and the learning algorithm of the BEL model are introduced in section 2. Moreover, the stability analysis of the BEL learning weights is given in section 2. In section 3, the BEL-based intelligent control scheme of the UAV is built up and the control simulation is carried out in section 5. Finally, conclusions are drawn in section 6.

2. MODELLING OF A SMALL UAV

A physical picture of the UAV studied in this work is shown in Figure 1. It is a small screw propeller UAV of a type usually used in reconnaissance.

For the small UAV without the thrust deflection of an engine, the mathematical model is in the same form as a traditional UAV model, which includes dynamics equations and kinemics equations. The aerodynamic parameters of the small UAV are obtained by blowing experiments.

The components of the total external force on airframe coordinates are shown as

$$\begin{bmatrix} F_X \\ F_Y \\ F_Z \end{bmatrix} = \begin{bmatrix} T \\ 0 \\ 0 \end{bmatrix} + S_{ba} \begin{bmatrix} -C_D QS \\ C_Y QS \\ -C_L QS \end{bmatrix} + \begin{bmatrix} G_X \\ G_Y \\ G_Z \end{bmatrix} \tag{1}$$

where T is thrust, G_X, G_Y, G_Z are the components of the gravity on the three axes of aircraft-body coordinate frame, C_D, C_Y, C_L are respectively coefficients of total aerodynamics on the three axes X_a,

Y_a, Z_a of the airflow coordinate Q is freedom flow pressure, S is wing area and S_{ba} is the transformation matrix from the airflow coordinate to aircraft-body coordinate frame.

Figure 1. UAV physical diagram

The gravity vector does not generate the torque because it passes through the centroid, then total external torque of normal UAV without thrust deflection is only the aerodynamic torque, and its components on the aircraft-body coordinate frame are described as

$$\begin{bmatrix} \bar{L} \\ M \\ N \end{bmatrix} = \begin{bmatrix} C_l QSb \\ C_m QSc \\ C_n QSb \end{bmatrix} \tag{2}$$

where C_l, C_m, C_n are the aerodynamic torque coefficients around the X_b, Y_b, Z_b axes on the aircraft-body coordinate frame.

Therefore, it is easy to obtain the dynamics equations of the small UAV which are represented as

$$\begin{bmatrix} \dot{V} \\ \dot{\alpha} \\ \dot{\beta} \end{bmatrix} = \begin{bmatrix} \dfrac{F_x \cos\alpha\cos\beta + F_y \sin\beta + F_z \sin\alpha\cos\beta}{m} \\ \dfrac{\frac{F_z}{m}\cos\alpha - Vp\cos\alpha\sin\beta + V\cos\beta q - \frac{F_x}{m}\sin\alpha - Vr\sin\alpha\sin\beta}{V\cos\beta} \\ \dfrac{\frac{F_y\cos^2\beta - F_x\cos\alpha\sin\beta\cos\beta - F_z\sin\alpha\sin\beta\cos\beta}{m} - Vr\cos\alpha\cos\beta + Vp\sin\alpha\cos\beta}{V\cos\beta} \end{bmatrix}$$

(3)

$$\begin{cases} \dot{p} = \dfrac{I_z\overline{L} + I_{zx}N + [I_z(I_y - I_z) - I_{zx}{}^2]qr + [(I_x - I_y + I_z)I_{zx}]pq}{I_x I_z - I_{zx}{}^2} \\ \dot{q} = \dfrac{M - I_{zx}(p^2 - r^2) - (I_x - I_z)pr}{I_y} \\ \dot{r} = \dfrac{I_{zx}\overline{L} + I_x N + (I_{zx}{}^2 - I_x I_y + I_x{}^2)pq - (I_x I_{zx} - I_y I_{zx} + I_z I_{zx})qr}{I_x I_z - I_{zx}{}^2} \end{cases} \quad (4)$$

where V is the flight speed, α is the attack angle, β is the sideslip angle, G_{Xa}, G_{Ya}, G_{Za} are the components of the gravity on the three axes of airflow coordination, p, q, r are the components of the angular rate on the aircraft-body coordinate frame and I is the rotational inertia (the subscript is the rotational axis). The kinemics equations are given by

$$\begin{bmatrix} \dot{\phi} \\ \dot{\theta} \\ \dot{\psi} \end{bmatrix} = \begin{bmatrix} p + (r\cos\phi + q\sin\phi)\tan\theta \\ q\cos\phi - r\sin\phi \\ \dfrac{r\cos\phi + q\sin\phi}{\cos\theta} \end{bmatrix} \quad (5)$$

$$\begin{bmatrix} \dot{x} \\ \dot{y} \\ \dot{z} \end{bmatrix} = \begin{bmatrix} \cos\theta\cos\psi & \cos\theta\sin\psi & -\sin\theta \\ \sin\theta\cos\psi\sin\phi - \sin\psi\cos\phi & \sin\theta\sin\psi\sin\phi + \cos\psi\cos\phi & \cos\theta\sin\phi \\ \sin\theta\cos\psi\cos\phi + \sin\psi\sin\phi & \sin\theta\sin\psi\cos\phi - \cos\psi\sin\phi & \cos\theta\cos\phi \end{bmatrix}^T \begin{bmatrix} u \\ v \\ w \end{bmatrix} \quad (6)$$

3. MODELLING AND ANALYSING OF THE BEL ALGORITHM

Description of the BEL model

The structure diagram of the computational model of the BEL is shown in Figure 2. There are two input channels including sensory input (which is sent to the thalamus and sensory cortex) and an emotional cue (which is sent to the amygdala and orbitofrotal cortex as reinforcing signals). The amygdala receives inputs from the thalamus and sensory cortical areas, while the orbitofrontal cortex receives inputs from sensory cortical areas and the amygdala only. There is one node for every stimulus both in the amygdala and orbitofrontal cortex parts. In common, there is only one output of the BEL model.

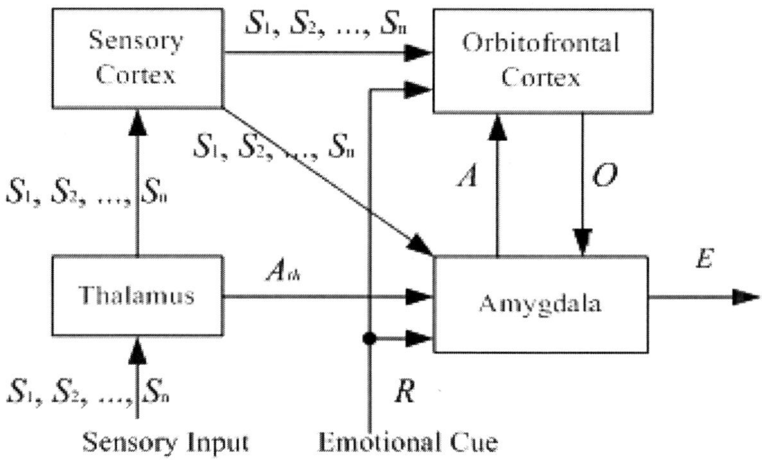

Figure 2. Structure scheme of the BEL model

As can be seen from Figure 2, the BEL system receives inputs from the sensory input S=[S1, S2, ⋯, Sn] and the emotional cue R, and only has one node for output signal E. Generally, the nodes in the orbitofrontal part are equal to the signal input number, while the nodes in the amygdala part are one more than in the orbitofrontal part.

According to the inner structure of the BEL model, the output of the amygdala and the output of the orbitofrontal are respectively described as

$$A = \sum_{i=1}^{n} S_i V_i + A_{th} V_{n+1} \tag{7}$$

$$O = \sum_{i=1}^{n} S_i W_i \tag{8}$$

where $V = [V_1, V_2, \cdots, V_{n+1}]^T$ and $W = [W_1, W_2, \cdots, W_n]^T$ are respectively the weights of nodes in the amygdala and the orbitofrontal parts, $A_{th} = \max\{S\}$ ..

The output of the BEL model is expressed by

$$E = A - O \tag{9}$$

As an intelligent model simulating the mammalian brain emotional learning mechanism, there must be a self-adaptive and self-adjusting part. Therefore, there are weights in the BEL system that play the role of such self-learning functions, through the adjusting steps named learning rates of the weights, given by

$$\Delta V_i = \alpha_A S_i \max(0, R - A),$$
$$\Delta V_{n+1} = \alpha_A A_{th} \max(0, R - A) \tag{10}$$

$$\Delta W_i = \alpha_O S_i (E - A_{th} V_{n-1} - R) \tag{11}$$

where α_A and α_O are the learning rate coefficients.

As can been seen from the BEL intelligent algorithm, we note that the weight-adjusting rule is monotonic due to the term $\max(\cdot,\cdot)$, the emotional reaction remains permanent if it begins learning, the duty

$$W(k) = \begin{bmatrix} W_1(k) & W_2(k) & \cdots & W_n(k) \end{bmatrix}^T \tag{19}$$

Then the state equation of W(k) are

$$W(k+1) = G(k)W(k) + H(k) \tag{20}$$

where

$$G(k) = I - \alpha_O S(k) S^T(k)$$

$$H(k) = -\alpha_O S(k)\left\{(R(k) - \begin{bmatrix} S^T(k) & 0 \end{bmatrix} V(k)\right\}$$

$$S(k) = \begin{bmatrix} S_1(k) & S_2(k) & \cdots & S_n(k) \end{bmatrix}^T$$

Assume at least one sensory input signal is no-zero, the two eigenvalues of G(k) are

$$\lambda_1 = 1, \quad \lambda_2 = 1 - \alpha_O(\sum_{i=1}^{n} S_i^2) \tag{21}$$

Theorem 2: Assume the reward signal and the sensory signal satisfies $R(k) = \begin{bmatrix} S^T(k) & 0 \end{bmatrix} V(k)$, and αO satisfies

$$0 \le \alpha_O \le \frac{2}{\sum_{i=1}^{n} \max\left\{S_i^2(k)\right\}} \tag{22}$$

Then system (20) is stable in the sense of Lyapunov.

Proof: Time-varying matrix G(k) is a real symmetric matrix, then we get

$$G(k) = U^T(k)\Lambda(k)U(k) \tag{23}$$

where U(k) is an orthogonal matrix, and

$$\Lambda(k) = diag\left\{1,1,\cdots,1,1-\alpha_O\left[\sum_{i=1}^{n} S_i^2(k)\right]\right\} \tag{24}$$

Select a positive Lyapunov function

$$L(k) = W^T(k)W(k) \tag{25}$$

According to (23), and R=0, we get

$$\begin{aligned}\Delta L(k) &= W^T(k)\left[G^2(k) - I\right]W(k)\\ &= \left[U(k)W(k)\right]^T\left[\Lambda^2(k) - I\right]\left[U(k)W(k)\right]\end{aligned} \tag{26}$$

where

$$\Lambda^2(k) - I = diag\left\{0,0,\cdots,0,\left[1-\alpha_O\sum_{i=1}^{n} S_i^2(k)\right]^2 - 1\right\}.$$

Therefore, if satisfying (22), we get $\Delta L(k) < 0$, therefore system (20) is stable in the sense of Lyapunov.

4. ATTITUDE CONTROL SYSTEM OF A UAV BASED ON BEL

The nonlinear model of the UAV in this work is expressed by

$$\dot{X} = f(X,U) \tag{27}$$

where $X=[u,v,w,p,q,r,\theta,\psi,\phi,x,y,h]^T$ is the state vector, in which u,v,w are the components of the flight velocity on the body coordinate, p,q,r are the components of the angular rate on the body coordinate, x,y,h are the position of the aircraft particle on the Earth coordinate and θ,ψ,ϕ are the attitude angles. $U=[\delta_T, \delta_e, \delta_a, \delta_r]^T$ is the control vector, in which δT is the throttle (here it is constant) and δe,δa,δr are the deflection angles of the control surfaces.

The attitude control system of the UAV based on BEL is shown in Figure 3, where θg represents the desired pitch angle, ϕg represents the desired roll angle and ψg represents the desired pitch angle.

The control laws in the elevator controller, aileron controller and rudder controller are respectively designed as

$$\begin{aligned}\Delta\delta_e &= K_{\Delta\theta}(\Delta\theta_g - \Delta\theta) + K_{\Delta q}\Delta q \\ &+V_3^{(ele)} \max\{(\Delta\theta_g - \Delta\theta), \Delta q\}\end{aligned} \tag{28}$$

$$\delta_a = K_\phi(\phi_g - \phi) + K_p p + V_3^{(ail)} \max\{\phi_g - \phi, p\} \tag{29}$$

$$\delta_r = K_\psi(\psi_g - \psi) - K_r r + V_3^{(rud)} \max\{\psi_g - \psi, r\} \tag{30}$$

where {K∗} are control gains, V3(*) are the weights of nodes for the signal from the thalamus in the corresponding BEL algorithm.

The BEL intelligent algorithm is used to on-line regulate the control gains. The BEL algorithm for the elevator controller is designed by

$$S^{(ele)} = \left[\Delta\theta_g - \Delta\theta,\ \Delta q\right] \tag{31}$$

$$R^{(ele)} = \lambda_{\Delta\theta_g}(\Delta\theta_g - \Delta\theta) + \lambda_{\Delta q}\Delta q \tag{32}$$

$$\begin{bmatrix} \Delta K_{\Delta\theta} \\ \Delta K_{\Delta q} \end{bmatrix} = \begin{bmatrix} \Delta V_1^{(ele)} - \Delta W_1^{(ele)} \\ \Delta V_2^{(ele)} - \Delta W_2^{(ele)} \end{bmatrix} \tag{33}$$

The BEL algorithm for the aileron controller is designed by

$$S^{(ail)} = \left[\phi_g - \phi,\ p\right] \tag{34}$$

$$R^{(ail)} = \lambda_{\phi_g}(\phi_g - \phi) + \lambda_p p \tag{35}$$

$$\begin{bmatrix} \Delta K_\phi \\ \Delta K_p \end{bmatrix} = \begin{bmatrix} \Delta V_1^{(ail)} - \Delta W_1^{(ail)} \\ \Delta V_2^{(ail)} - \Delta W_2^{(ail)} \end{bmatrix} \quad (36)$$

The BEL algorithm for the rudder controller is designed by

$$S^{(rud)} = \left[\gamma_1(\psi_g - \psi),\ \gamma_2 r \right] \quad (37)$$

$$R^{(rud)} = \lambda_{\psi_g}(\psi_g - \psi) + \lambda_r r \quad (38)$$

$$\begin{bmatrix} \Delta K_\psi \\ \Delta K_r \end{bmatrix} = \begin{bmatrix} \Delta V_1^{(rud)} - \Delta W_1^{(rud)} \\ \Delta V_2^{(rud)} - \Delta W_2^{(rud)} \end{bmatrix} \quad (39)$$

where ΔV_1, ΔV_2, ΔV_3, ΔW_1 and ΔW_2 are the learning rates of BEL systems. γ_1 and γ_2 are constant coefficients. It should be mentioned that these two coefficients can also exist in (31) and (34).

From the above equations we can see that the BEL-based intelligent controllers almost have the same forms as the traditional controllers for attitude control of the UAV. The main difference between them is that the BEL intelligent controllers can regulate the control gains on-line, which makes them adaptive to variations in system conditions.

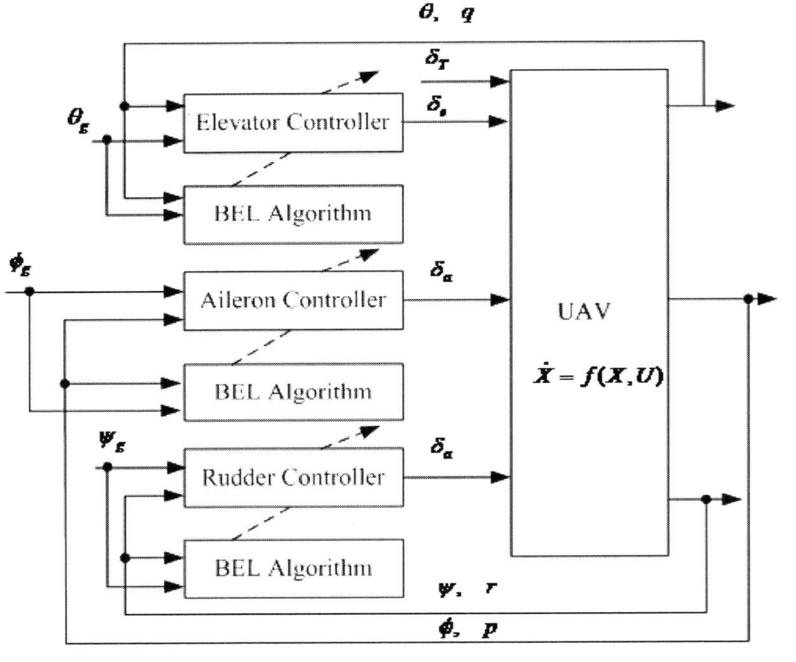

Figure 3. BEL-based flight control system of the small UAV

5. CONTROL SIMULATION ANALYSIS

The main purpose of this work is try to use a structural model based on the limbic system of the mammalian brain, for the attitude stability control of a small UAV with wind interference, to investigate the attitude keeping capacity of the BEL-based intelligent controller.

First, we use the traditional control scheme, while the control gains can be obtained using the single-loop design method, the values of which are shown in the following:

$$K_{\Delta\theta} = 1.8 \text{ , } K_{\Delta q} = 0.8$$

$$K_{\phi} = 0.5 \text{ , } K_{p} = 0.1$$

$$K_{\psi} = 0.5 \text{ , } K_{r} = 0.1$$

Then, for justice, the above control gains are also used in the BEL intelligent controller, as the initial control gains.

Suppose the small UAV is in flat flight, that is $\phi=0$, $p=q=r=0$, $\phi^{\cdot}=\theta^{\cdot}=\psi^{\cdot}=0$, $u^{\cdot}=v^{\cdot}=w^{\cdot}=p^{\cdot}=q^{\cdot}=r^{\cdot}=0$ on the fixed line, and is disturbed by a wind from the west to the east, from top to bottom, the angle on the horizontal plane is 60 deg and the wind speed is 18 m/s.

After calculating the equilibrium state of the nonlinear model of the UAV, the desired pitch angle θg can be obtained. Consider the control strategy described in Figure 3, in which the structure of the BEL model is described in Figure 2.

Figure 4~Figure 6 respectively show the system responses of the pitch attitude angle, roll attitude angle and yaw attitude angle based on the traditional controller and the BEL intelligent controller. From the simulation results, we can ascertain that the BEL intelligent controller makes the flight attitudes return to the balanced states quicker and with higher precision than the traditional controller, which shows that the BEL intelligent controller has stronger anti-disturbance capability than the traditional controller.

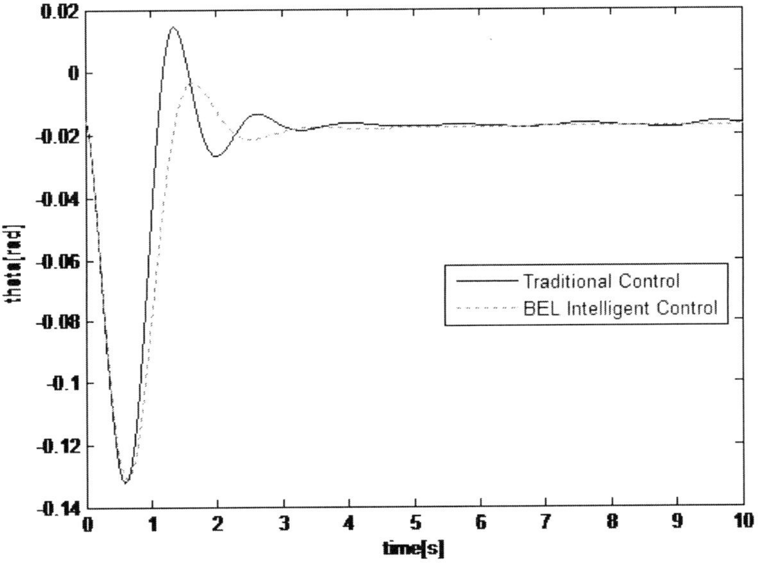

Figure 4. Pitch angle responses

Figure 7~Figure 9 respectively give the elevator deflection angle curses, aileron deflection angle curses and rudder deflection angle curses of the traditional and BEL intelligent controllers. According to the simulation results, for the BEL intelligent controller, the control outputs of the elevator and rudder are smaller than those of the traditional controller, and the control output of the aileron is the opposite. This shows that the control gains of the BEL intelligent controller are learning and adjusting through the control progress, achieving a better performance, while the parameters of the traditional controller keep constant.

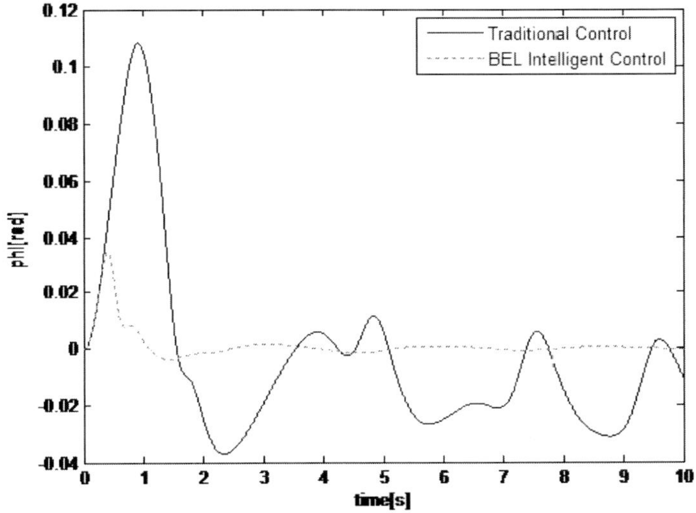

Figure 5. Roll angular responses

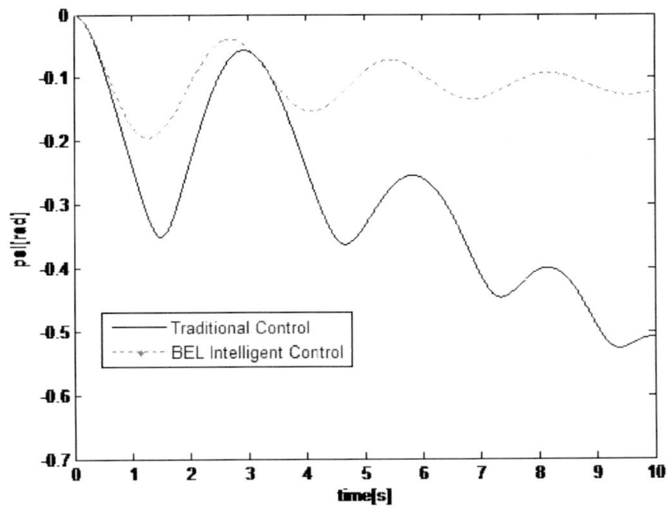

Figure 6. Yaw angle responses

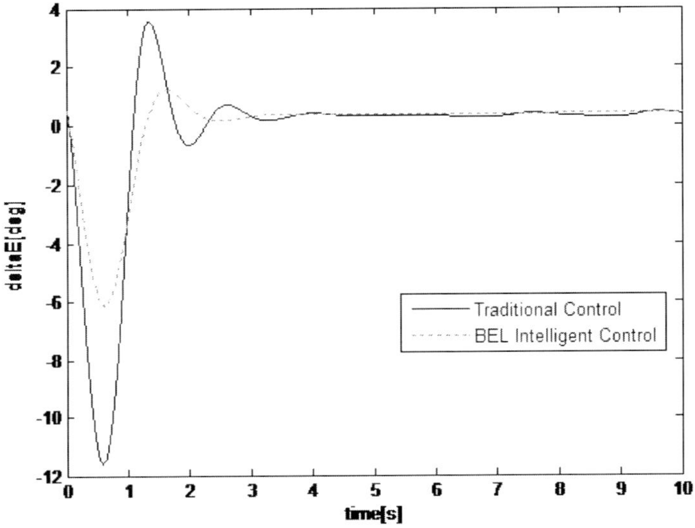

Figure 7. Elevator control output

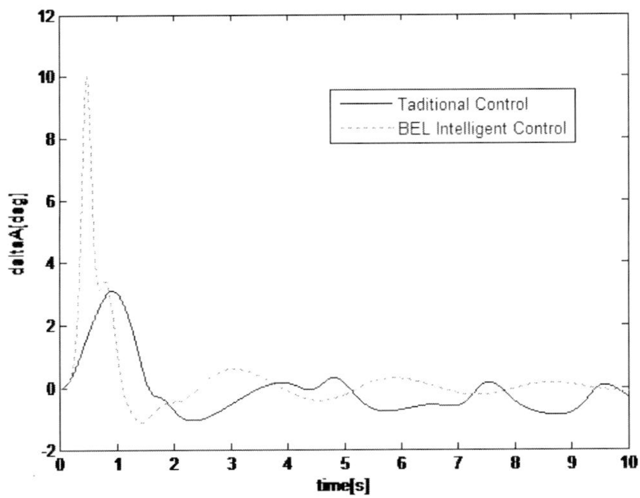

Figure 8. Aileron control output

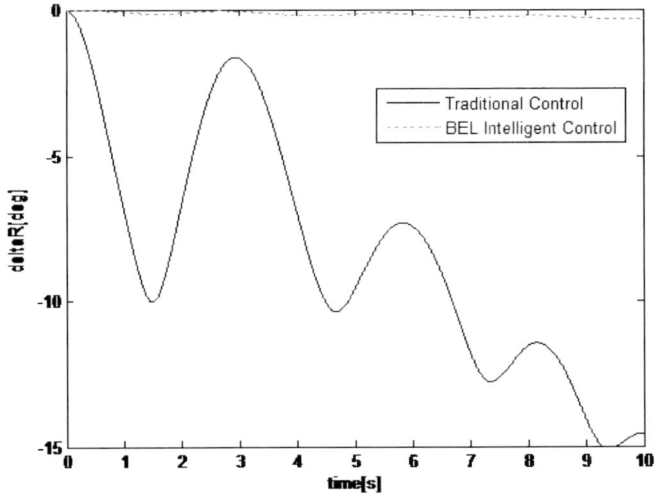

Figure 9. Rudder control output

6. CONCLUSIONS

To counter the difficulties in the attitude control of small UAVs, a new intelligent controller which is based on the BEL model inspired by the emotional learning process in the A-O system in mammalian brains is applied. The stability of the learning algorithm of BEL is initially investigated in the sense of Lyapunov. Three BEL models are used in designing the control rules of the three manipulators including elevator, aileron and rudder, actually adjusting the control gains of the three controllers.

In the simulation, the BEL intelligent control is applied in the attitude control loop of the UAV under the condition of wind disturbance, making the UAV stable and recovering to the flat flight state. Simulation results show the better control precision and stronger robustness of the BEL intelligent control system, when compared with the traditional flight control system.

However, the selections of the feedback signal, the function forms and the weight coefficients of the sensory input and the emotional cue are the main problem of BEL intelligent control.

In conclusion, BEL intelligent control can be used in a real-time control system on account of the small computational cost and adaptive capacity. Therefore, the promising potential in the application in control fields can be affirmed.

7. ACKNOWLEDGEMENTS

This study was supported by the NUAA Fundamental Research Funds (no. xlink013029, NP2011012, NP2012101), Aeronautical Science Foundation of China (no. 2010ZA52002), and Specialized Research Fund for the Doctoral Program of Higher Education (no. 20123218120015).

REFERENCES

1 Wei Ren. "Trajectory tracking control for miniature fixed-wing unmanned air vehicle". *International Journal of System Science*. 2007, 38(4), 361-368.

2 Wei Ren. "On constrained nonlinear tracking control of a small fixed-wing UAV". *Journal of Intelligent and Robotic Systems*. 2007, 48(4), 525-537.

3 Campa G., Gu Y., Seanor B., et al. "Design and flight-testing of non-linear formation control laws". *Control Engineering Practice*, 2007, 15: 1077-1092.

4 Moren J., Balkenius C. "A computational model of emotional learning in the amygdala." From Animals to Animals 6: Proceedings of the 6th International Conference on the Simulation of Adaptive Behavior, Cambridge, Mass, The MIT Press, 2000.

5 Moren J. "Emotion and learning-A computational model of the amygdala." [PhD Dissertation], England: Lund University. 2002.

6 Lucas C., Shahmirzadi D., Sheikholeslami N. "Introducing BELBIC: Brain emotional learning based intelligent controller." *International Journal of Intelligent Automation and Soft Computing*, 2004, 10(1): 11-22.

7 Jamali M. R., Arami1 A., Hosseini B., Moshiri1 B. and Lucas C. Real time emotional control for anti-swing and positioning control of simo overhead traveling crane. *International Journal of Innovative Computing, Information and Control*, 2008, 2008, Volume 4, Number 9, September pp. 2333—2344.

8 Jamali M.R., Arami A., Dehyadegari M., Lucas C., Navabi Z. Emotion on FPGA: Model driven approach, Expert Systems with Applications 36 (2009) 7369–7378.

9 Rouhani H., Sadeghzadeh A., Lucas C. and Nadjar Araabi B. Emotional learning based intelligent speed and position control applied to neurofuzzy model of switched reluctance motor. *Control and Cybernetics*, vol. 36 (2007) No. 1,75-93

10 Masoudinejad M., Khorsandi R., Fatehi A., Lucas C., Fakhimi Derakhshan S., Jamali M. R.. Real-Time Level Plant Control Using Improved BELBIC, Proceedings of the 17th World Congress The

International Federation of Automatic Control, Seoul, Korea, July 6-11, 2008.

11 Babaie T., Karimizandi R. and Lucas C. Learning based brain emotional intelligence as a new aspect for development of an alarm system, *Soft Computing* (2008) 12:857–873

12 Rouhani H., Jalili M., Araabi B.N., Eppler W., Lucas C. (2007). “Brain emotional learning based intelligent controller applied to neurofuzzy model of micro-heat exchanger.” *Expert Systems with Applications*, 2007, 32: 911-918.

13 Ziyang Zhen, Daobo Wang, Zhisheng Wang. “Flight simulation servo system based on learning based intelligent controller.” Proceedings of the Fourth International Conference on Impulse and Hybrid Dynamic Systems. Nanning, Guangxi, China, July 20-22, 2007: 1372-1375.

14 Yong Hu, Ziyang Zhen, Zhisheng Wang, Mingzhi Geng. “Brain emotional learning based adaptive identification method for nonlinear dynamic system.” 2008 Chinese Control and Decision Conference, Yantai, Shandong, China, July 2-4, 2008: 3498-3501.

15 Mehrabian A.R. and Lucas C. “Emotional learning based intelligent robust adaptive controller for stable uncertain nonlinear systems.” *International Journal of Computational Intelligence*, 2005, 2(4): 1304-4508.

CHAPTER 7

SAMPLING-BASED REAL-TIME MOTION PLANNING UNDER STATE UNCERTAINTY FOR AUTONOMOUS MICRO-AERIAL VEHICLES IN GPS-DENIED ENVIRONMENTS

Dachuan Li [1,*], Qing Li [1], Nong Cheng [1,2] and Jingyan Song [1]

[1] Department of Automation, Tsinghua University, Bejing 100084, China

[2] National Key Laboratory on Flight Vehicle Control Integrated Technology, Flight Automatic Control Research Institute, Xi'an 710065, China

ABSTRACT

This paper presents a real-time motion planning approach for autonomous vehicles with complex dynamics and state uncertainty. The approach is motivated by the motion planning problem for autonomous vehicles navigating in GPS-denied dynamic environments, which involves non-linear and/or non-holonomic vehicle dynamics, incomplete state estimates, and constraints imposed by uncertain and cluttered environments. To address the above motion planning problem, we propose an extension of the closed-loop rapid belief trees, the closed-loop random belief trees (CL-RBT), which incorporates predictions of the position estimation uncertainty, using a factored form of the covariance provided by the Kalman filter-based estimator. The proposed motion planner operates by incrementally constructing a tree of dynamically feasible trajectories using the closed-loop prediction, while selecting candidate paths with low uncertainty using efficient covariance update and propagation. The algorithm can operate in real-time, continuously providing the controller with feasible paths for execution, enabling the vehicle to account for dynamic and uncertain environments. Simulation results demonstrate that the proposed approach can generate feasible trajectories that reduce the state estimation uncertainty, while handling complex vehicle dynamics and environment constraints.

KEYWORDS

motion planning; micro-aerial vehicles; rapidly exploring random trees (RRT); state estimation uncertainty

1. INTRODUCTION

Autonomous micro-aerial vehicles (MAV) are playing an increasingly important role in many civil and military applications. MAVs that are capable of autonomously making decisions and operating can be applied in many tasks scenarios that are not accessible by humans and ground mobile robots, such as indoor exploration and mapping, search and rescue, disaster relief, *etc.* As a result, there has been an increasing interest in developing autonomous MAV systems for indoor navigation. In recent years, many researchers have developed and implemented various kinds of MAV systems demonstrating various degrees of autonomy in performing tasks, such as indoor exploration [1], environment mapping [2] and agile flight [3].

Despite the considerable progress achieved in this domain, there are still many challenges in developing fully autonomous MAV systems. One of the key problems is the motion/path planning for MAVs capable of operating in GPS-denied complex environments. For MAVs performing tasks in such scenarios, the motion planning algorithm must comply with constraints, including complex vehicle dynamics (non-linear and/or non-holonomic dynamics with a high dimensional state space) and environmental constraints (unstructured and uncertain, time-varying operating environments). Particularly, since MAVs typically cannot directly get access to the information of the current state, they must estimate the distribution over the states using measurements from onboard sensors or external aids (GPS [4], external motion capture system [5]). However, GPS is unreliable in urban canyons and completely unavailable in most indoor environments; the external motion capture system is also impractical for such scenarios, since it requires pre-installation of camera arrays in the environment. As a result, the state estimation of MAVs must rely only on onboard sensing capabilities, which is highly constrained in terms of precision and range due to the size and weight constraints of the MAVs. Moreover, the performance of MAV state estimation and localization using exteroceptive sensors (laser rangefinder [6], camera [7] and RGB-D sensor [8]) varies across the environment depending on the distinctive features and perceptual structure of the environments. As a result, the motion planning progress must also integrate the uncertainty of state estimation to ensure reliability and robustness to imperfect and noisy state estimates caused by limited sensing capability. These challenges require that the motion/path planner must be able to generate feasible paths that satisfy the constraints imposed by uncertain and unstructured environments, as well as complex vehicle dynamics, while ensuring measurement gathering along the path to reduce state estimation uncertainty.

In this paper, we propose a real-time motion planning strategy (closed-loop random belief trees, CL-RBT) for autonomous vehicles with complex dynamics and in the presence of state estimation uncertainty, while handling the constraints imposed by the uncertain and dynamic operating environments. The proposed motion planning strategy is built upon the randomized sampling-based motion framework, the real-time closed-loop rapidly exploring random trees (CL-RRT) [9], which operates by sampling in the input space of the controller and generates trajectories through forward closed-loop simulation using the vehicle dynamics model and path-tracking controller. This allows the motion planner to easily account for non-linear and non-holonomic vehicle dynamics and dynamic uncertain environments. We extend the general CL-RRT to handle the state uncertainty by incorporating the prediction of posterior state estimation uncertainty in the motion planning framework. The uncertainty of state estimation is characterized using a factored form of the covariance [10] provided by the Kalman filter class-based estimator. This factored form enables efficient covariance propagation in that it combines the update of covariance along a path from multi-step observations into a single linear step, and the posterior covariance of a certain path after adding new nodes can also be updated online. The overall motion planning operates by incrementally constructing a tree of dynamically feasible trajectories in real time, while continuously selecting and executing trajectories that are a tradeoff between minimizing path cost and reducing localization uncertainty. We validated the proposed algorithm in various illustrative scenarios of a simulated quadrotor MAV with non-linear dynamics and limited sensing in enclosed and unstructured GPS-denied environments. Simulation results demonstrate that the CL-RBT can efficiently generate dynamically feasible trajectories that ensure state estimation accuracy, while preserving the property of the CL-RRT framework. The CL-RBT can handle complex vehicle dynamics and state uncertainty, enabling the MAV to autonomously navigate in uncertain, unstructured and GPS-denied environments.

The paper is organized as follows: Section 2 reviews related work on motion planning under uncertainty. Section 3 presents the formulation of the motion planning problem. Section 4 provides the covariance propagation approach. A detailed description of the CL-RBT is presented in Section 5, followed by the simulation results and analysis in Section 6. Finally, the paper is concluded in Section 7, with a discussion on future work.

2. RELATED WORK

Conventionally, motion planning approaches under state uncertainty are typically formulated as a partially observable Markov decision process (POMDP) problem [11], which provides the most general mathematical framework for solving the planning problem with partial observability. While POMDP has been applied to low-dimensional and small-scale problems [12], most POMDP-based approaches rely on discretizing the state space, making it computationally intractable for realistic applications. Recently, many approaches have been proposed to address the problem of scalability using approximation and iterative methods. Van den Berg *et al.* [13] proposed an iterative motion planning approach to solve the continuous POMDP problem, using a belief space variant of the iterative LQG (linear-quadratic Gaussian) method. Similarly, Bai *et al.* [14] also formulated the problem as a continuous POMDP and solved the problem with a Monte Carlo value iteration method. However, these approaches are still less effective at addressing problems with complex vehicle dynamics and motion planning with large-scale, high dimensional state space or configuration space.

In contrast, sampling-based motion planning strategies have been widely acknowledged as effective approaches for solving motion planning problems with high dimensional configuration space and complex vehicle dynamics. In particular, the PRM (probabilistic roadmap) [15] and RRT (rapidly exploring random trees) [16] and their variants are the most widely applied sampling-based approaches, and they have been successfully applied in a number of applications. More recently, the closed-loop RRT [9,17,18] extends the conventional RRT by incorporating forward prediction using closed-loop dynamics model with controller. The CL-RRT can generate more feasible paths that account for complex vehicle dynamics, and it can be implemented in real-time to handle dynamic, uncertain environments. Several approaches have been proposed to incorporate state uncertainty in the sampling-based motion planning framework. Van den Berg *et al.* [19] proposed the LQG-MP strategy by integrating the *a priori* distribution over state estimates into the standard RRT framework. However, the LQG-MP algorithm is intractable for real-time applications, due to its high computation cost. In [20], Bry *et al.*, combined an LQG-based covariance pruning technique with the RRT* framework for planning in belief space. Alternative to RRT-based approaches, the belief roadmap (BRM) extends the PRM by integrating the predictions of state estimation uncertainty using an EKF (extended Kalman filter) estimator and performs belief planning by searching paths with the lowest uncertainty from the roadmap. The BRM was later extended to use the UKF estimator and heuristic sampling strategy [21], and it was further implemented on a quadrotor and proven successful for indoor flight tests [22]. However, the BRM assumes that the vehicle is fully controllable and treats the problem as a simple kinematic motion planning problem, e.g., it does not consider complex vehicle dynamics and the feasibility of the generated paths. As a result, it is not yet well suited to vehicles with complex dynamic constraints and dynamic environments. In addition, the BRM operates by planning on pre-constructed static graphs, and therefore, it is not a real-time planning algorithm in essence. Recent research has also considered another similar kind of planning problem with uncertainty, where the planner must generate paths that maximize information collection and reduce the uncertainty on the location estimates of features or targets in the environment. The information-rich RRT (IRRT) [23] addresses this path planning problem by incorporating the qualification of information gain into the CL-RRT framework, using the Fisher information matrix. This kind of problem can be regarded as the inverse problem of planning with uncertainty on the vehicle's own state.

3. PROBLEM FORMULATION

3.1. Motion Planning under State Uncertainty

Consider a vehicle with non-linear dynamics that operates in an environment without external positioning systems. The vehicle typically does not have direct access to the accurate information of its current state, but instead obtains the estimates of the state using observations derived from the measurements of onboard sensors, which are typically noisy and incomplete. The stochastic dynamics and observation model of the system can be given by the following discrete time state-transition form:

$$\begin{aligned} \mathbf{x}_{t+1} &= f(\mathbf{x}_t, \mathbf{u}_t, \mathbf{w}_t) \qquad \mathbf{w}_t \sim N(0, \mathbf{Q}) \\ \mathbf{z}_t &= f(\mathbf{x}_t, \mathbf{v}_t) \qquad \mathbf{v}_t \sim N(0, \mathbf{R}) \end{aligned} \tag{1}$$

where $\mathbf{x}\in X$,$\mathbf{u}\in U \in$ and $\mathbf{z}\in Z$ are the vehicle's state, control input and observation, respectively. $\mathbf{w}_t$ and $\mathbf{v}_t$ denote the process disturbance and measurement noise, and both can be formulated as Gaussian noise with zero mean and covariance ($\mathbf{Q}$ and$\mathbf{R}$, respectively). Assuming that all probability distributions are Gaussian, given the previous observation ($\mathbf{z}_{1:t}$) and control inputs ($\mathbf{u}_{1:t}$), the state is typically estimated using a Bayesian filter, providing the distribution of the vehicle's state (or the belief of the state):

$$p(\mathbf{x}_t) = p(\mathbf{x}_t \mid \mathbf{u}_{1:t}, \mathbf{z}_{1:t}) \tag{2}$$

which is characterized by a mean state μ_t and a covariance $\Sigma_t(b(\mathbf{x}_t) = (\mu_t, \Sigma_t))$. The mean μ_t provided by the Bayesian filter is a optimal estimation of the vehicle's state distribution, and it can be used for control and decision making, while the mean Σ_t characterizes the confidence (or uncertainty) of state estimation.

Denoting X_{free} as the subset of all collision-free states, given the initial state $\mathbf{x}_0 \in X_{free}$ and the goal region $X_{goal} \subset X_{free}$, as well as the partial knowledge of the environment, then the primary objective of motion planning is to find the control policy$\mathbf{u}_{0:T} = \pi[b(\mathbf{x}_{0:T})]$ and corresponding sequence of states $\mathbf{x}_{0:T}$, such that the vehicle reaches the goal region in a finite time horizon by applying the control policy:

$$\mathbf{x}_T \in X_{goal}, t \in (0, t_f], t_f \in (0, \infty) \tag{3}$$

meanwhile minimizing the following objective function:

$$J(b(\mathbf{x}_t)) = E_{b(\mathbf{x}_{goal})|b(\mathbf{x}_t),\mathbf{u}_{0:T}}[C(\mathbf{x}_t \mid \mathbf{x}_{goal})] + \sum_{t=0}^{T} C(\mathbf{x}_t, \mathbf{u}_t) \tag{4}$$

where $C(\mathbf{x}_t|\mathbf{x}_{goal})$ is the expected cost function from $\mathbf{x}_t$ to $\mathbf{x}_{goal}$; note that it takes the expectation form, since the measurement and state are both probabilistic, and $C(\mathbf{x}_t, \mathbf{u}_t)$ is the cost by applying control $\mathbf{u}_t$. The generated trajectory must also satisfy the constraints imposed by the vehicle dynamics Equation (1) and environments (*i.e.*, collision avoidance, $\mathbf{x}_{0:T} \in X_{free}$). Moreover, when the vehicle does not have accurate knowledge of the estimate of its state, the motion planner must take the uncertainty of the state estimate along the generated paths into account, thus the motion planning problem into the belief planning problem (*i.e.*, planning in belief space [20]). By integrating the information from the mean and covariance of the belief, the motion planning algorithm must choose actions and beliefs, such that the posterior covariance at the end of the trajectory is minimized, yielding trajectories that achieve a trade-off between low path cost and high confidence in the state estimation.

3.2. State Estimation of Gaussian Systems

Consider a system with the dynamic and observation model in the form of Equation (1), one of the most common and robust methods for estimating the distribution over its state is the Bayesian filtering [10]. Given knowledge of the prior control input $\mathbf{u}_t$ and sensor measurement $\mathbf{z}_{t+1}$, the belief of the state $b(\mathbf{x}_{t+1})$ after a sequence of control and observation can be estimated as:

$$b(\mathbf{x}_{t+1}) = p(\mathbf{x}_{t+1} \mid \mathbf{u}_{1:t}, \mathbf{z}_{1:t}) = \lambda p(\mathbf{z}_{t+1} \mid \mathbf{x}_{t+1}) \int p(\mathbf{x}_{t+1} \mid \mathbf{x}_t, \mathbf{u}_t) b(\mathbf{x}_t) d\mathbf{x}_t \tag{5}$$

where λ is a normalization factor.

Assuming the state and observation transition function (f, h) are linear with state and observation, which are both Gaussian distributions; the Bayesian filtering can be implemented as the Kalman filter [24], while the extended Kalman filter (EKF) [25] is typically applied for estimating the state distribution of the system with nonlinear state and observation transition functions. For a system model of the form of Equation (1), the EKF state estimation can be divided into the process step and measurement update step. Denoting the distribution of the state $b(\mathbf{x}_t) = N(\hat{\mathbf{x}}_t, \Sigma_t)$, the process step first predicts the state using the control and state of the previous step, which is given as:

$$\begin{aligned} \bar{\hat{\mathbf{x}}}_t &= g(\hat{\mathbf{x}}_{t-1}, \mathbf{u}_t) \\ \bar{\Sigma}_t &= G_t \Sigma_{t-1} G_t^T + V_t Q_t V_t^T \end{aligned} \tag{6}$$

where G_t is the Jacobian of g with respect to the state, and V_t is the Jacobian of g with respect to $\mathbf{w}_t$. Then, the measurement step adjusts the estimate and covariance by incorporating the information from new observations:

$$\begin{aligned}\hat{\mathbf{x}}_t &= \bar{\hat{\mathbf{x}}}_t + K_t(\mathbf{z}_t - H_t\bar{\hat{\mathbf{x}}}_t) \\ \Sigma_t &= \bar{\Sigma}_t - K_t H_t \bar{\Sigma}_t\end{aligned} \tag{7}$$

where H_t denotes the Jacobian of h with respect to the state, and K_t is the Kalman gain, which is updated by:

$$K_t = \bar{\Sigma}_t H_t^T (R_t + H_t \bar{\Sigma}_t H_t^T)^{-1} \tag{8}$$

Typically, the control and decision making are based on the mean $\hat{\mathbf{x}}_t$ of the estimated distribution provided by the EKF, and the covariance, while the covariance Σ_t captures the uncertainty of the state estimate using the sensor measurements. Therefore, the uncertainty of the state estimate along the trajectory from the path planner can be evaluated using the norm of the covariance, which will be discussed in the next section.

4. LINEAR COVARIANCE PROPAGATION

In order to evaluate the state uncertainty resulting from a specific planned trajectory, the most common approach for a system with the EKF estimator is to compute the posterior covariance of the ending belief $b(\mathbf{x}_t)$ from the sequence of actions and measurements along the trajectory. However, the propagation of the posterior covariance requires multiple iterative calculations of the EKF process and measurement updating from the initial belief according to Equation (4), leading to a heavy computational cost. This is even worse when the initial belief is modified, since the posterior covariance must be re-computed using the entire EKF updates from the new initial belief. In particular, this fact has a more significant effect on a sampling-based motion planner, since it generally constructs a graph and tree of multiple trajectories, thus different paths can result in different posterior covariance to the same node, this requires that the motion planner must perform the propagation process for each covariance.

To reduce the computational cost of the covariance propagation, we rely on previous results on the factorization of the covariance [10], which allows the propagation of posterior covariance to be propagated in a single linear update step instead of multiple non-linear updates for the EKF filter.

Following Theorem 1 in [10], the factorization of covariance is given by:

$$\Sigma_t = \Lambda_t \Pi_t^{-1} \tag{9}$$

where Λ_t and Π_t can be calculated as linear functions of Λ_{t-1} and Π_{t-1}, using the EKF process and measurement update step.

The factorization of the covariance matrix can be proved using the following matrix inversion lemma:

Lemma 1. For matrices M_1, M_2, $M_3 \in \mathrm{R}^{n\times n}$, we have:

$$(M_1 + M_2 M_3^{-1})^{-1} = M_3 (M_2 + M_1 M_3)^{-1} \tag{10}$$

Denote the initial state covariance as Σ_0, and Σ_0 can be factored as:

$$\Sigma_0 = \Sigma_0 \mathbf{I}^{-1} \tag{11}$$

Given $\Sigma_{t-1} = \Lambda_{t-1}\Pi_{t-1}^{-1}$ and denoting $S_t = V_t Q_t V_t^T$, the process update of EKF (Equation (3)) can be written as:

$$\overline{\Sigma}_t = G_t \Sigma_{t-1} G_t^T + S_t = G_t \Lambda_{t-1} \Pi_{t-1}^{-1} G_t^T + S_t = G_t \Lambda_{t-1} (G_t^{-T} \Pi_{t-1})^{-1} + S_t$$

Following Equation (5), we have:

$$\overline{\Sigma}_t = \left((G_t^{-T} \Pi_{t-1})(G_t \Lambda_{t-1} + S_t G_t^{-T} \Pi_{t-1})^{-1} \right)^{-1} = (\overline{E}_t \overline{F}_t^{-1})^{-1} = \overline{F}_t \overline{E}_t^{-1} \tag{7}$$

where $\overline{E}_t = G_t^{-T} \Pi_{t-1}$ and $\overline{F}_t = G_t \Lambda_{t-1} + S_t G_t^{-T} \Pi_{t-1}$, For computation considerations, the covariance update of the measurement update step can be given by the information form [11]:

$$\begin{aligned} \overline{\Omega}_t &= \overline{\Sigma}_t^{-1} = (G_t \Omega_{t-1}^{-1} G_t^T + T_t)^{-1} \\ \Omega_t &= \overline{\Omega}_t + H_t^T R_t^{-1} H_t \end{aligned} \tag{8}$$

Denoting $N_t = H_t^T R_t^{-1} H_t$, Equation (8) can be rewritten as $\Omega_t = \overline{\Omega}_t + N_t$. Therefore, from Equations (7) and (8), we can write:

$\Sigma_t = \Omega_t^{-1} = (\bar{\Sigma}_t^{-1} + N_t)^{-1} = (\bar{E}_t \bar{F}_t^{-1} + N_t)^{-1}$ Following Equation (5),

$$\Sigma_t = \bar{F}_t(\bar{E}_t + N_t\bar{F}_t)^{-1} = \Lambda_t \Pi_t^{-1}$$

where:

$$\begin{aligned} \Lambda_t &= \bar{F}_t = G_t\Lambda_{t-1} + S_t G_t^{-T}\Pi_{t-1} \\ \Pi_t &= \bar{E}_t + N_t\bar{F}_t = G_t^{-T}\Pi_{t-1} + N_t(G_t\Lambda_{t-1} + S_t G_t^{-T}\Pi_{t-1}) \\ &= N_t G_t \Lambda_{t-1} + (I + N_t S_t)G_t^{-T}\Pi_{t-1} \end{aligned} \tag{9}$$

As can be seen from Equation (9), Λ_t, Π_t are both linear functions of Λ_{t-1}, Π_{t-1}.

Denoting Ψ as the stacked block matrix of Λ and Π:

$$\Psi = \begin{bmatrix} \Lambda \\ \Pi \end{bmatrix} \tag{10}$$

Equation (10) can be rewritten as:

$$\Psi_t = \begin{bmatrix} \Lambda_t \\ \Pi_t \end{bmatrix} = \begin{bmatrix} 0 & I \\ I & N \end{bmatrix}_t \begin{bmatrix} 0 & G^{-T} \\ G & SG^{-T} \end{bmatrix}_t \begin{bmatrix} \Lambda_{t-1} \\ \Pi_{t-1} \end{bmatrix} = \varsigma_t \Psi_{t-1} \tag{11}$$

where:

$$\varsigma_t = \begin{bmatrix} 0 & I \\ I & N \end{bmatrix}_t \begin{bmatrix} 0 & G^{-T} \\ G & SG^{-T} \end{bmatrix}_t = \begin{bmatrix} G & SG^{-T} \\ NG & G^{-T} + NSG^{-T} \end{bmatrix}_t \quad (12)$$

is the one-step transfer function matrix. Therefore, the covariance at t can be recovered using the factors of Ψ_t:

$$\Sigma_t = \Lambda_t \Pi_t^{-1} \quad (13)$$

Using the above factorization and the linear transfer function, the non-linear Kalman update process of the covariance can be transformed into a linear propagation step of the covariance factors, and the multiple updates can be combined into a single transfer step. This allows for efficient covariance propagation and uncertainty prediction for the sampling-based motion planning strategy: The posterior covariance Σ_T resulting from any initial belief (μ_t, Σ_t) along the trajectory can be recovered by multiplying multiple transfer functions:

$$\Psi_T = \varsigma_T \cdots \varsigma_{t+1} \Psi_t = \varsigma_{T-t} \Psi_t \quad (14)$$

and in a sampling-based path planner that represents paths in the form of trees and nodes, for any arbitrary node n_j in the trajectory, the posterior covariance of the state at the node can be propagated from the predecessor node *j–1* along the path in one efficient step using the transfer function $\varsigma_{0:j}$ (Figure 1):

$$\Psi_j = \varsigma_j \cdots \varsigma_1 \Psi_0 = \varsigma_{0:j} \Psi_0 \tag{15}$$

Taking advantage of the above linear covariance propagation approach, the state estimation covariance of one point n_i(node or time step) can be predicted using the covariance of its adjacent point, as well as the EKF Jacobian matrix set associated with the trajectory between n_{i-1} and n_i; this procedure can be given as follows:

Step 1:

Extract the covariance factors Λ_{i-1} and Π_{i-1} from block matrix Ψ_{i-1} of n_{i-1}.

Step 2:

Compute the EKF matrix set (S_i, G_i and N_i) based on the state transition model and observation model, as well as the predicted state and simulated observation (Equation (1)).

Step 3:

Compute the transfer function ς_i using S_i, G_i and N_i, according to Equation (12).

Step 4:

Propagate the block matrix Ψ_i of n_i based on ς_i and Ψ_{i-1}, according to Equation (11), then extract the covariance factors Λ_i and Π_i from block matrix Ψ_i.

Step 5:

Recover the covariance Σ_i of n_i using Λ_i and Π_i: $\Sigma_i = \Lambda_i \Pi_i^{-1}$.

This covariance propagation procedure allows the path planning algorithm to incorporate the covariance-based qualification of state estimation uncertainty into the planning progress in an efficient way. In the CL-RBT algorithm, which is described in detail in the following section, the state estimation uncertainty is qualified using the trace of the covariance:

$$J(\Sigma_i) = tr(\Sigma_i) \tag{16}$$

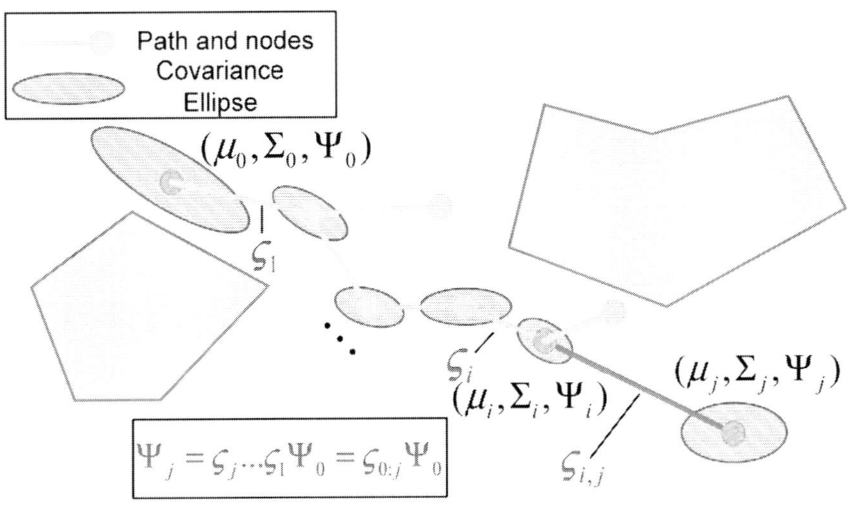

Figure 1. Linear covariance prediction procedure using the transfer function.

5. CLOSED-LOOP RANDOM BELIEF TREES ALGORITHM (CL-RBT)

5.1. Data Structure

Similar to the conventional RRT, the CL-RBT operates by constructing a tree of trajectories in the state space (more precisely, the belief space). The tree is defined by nodes and edges that connect different nodes. The primary information stored in a node n consists of the mean of the state, uncertainty qualification, the path cost and the parent node:

$$n_i \triangleq \{\mu_i, \Sigma_i, \Psi_i, J_i(\Sigma), C(n_i), n_i.\text{parent}\}$$

where μ_i and Σ_i represent the mean and the covariance of the state distribution, respectively. Ψ_i is the factored form of the covariance. $J_i(\Sigma)$ qualifies the uncertainty along the trajectory form the root node to node n_i, which can be calculated using the covariance propagation process described in Section 4: $J_i(\Sigma) = tr(\Sigma_i)$; $C(n_i)$ stores the data of the path execution cost associated with node n_i, which can be split into two parts:

$$C_i = \omega_1 c(n_i \mid n_{root}) + \omega_2 \hat{c}(x_{goal} \mid n_i)$$

where $c(n_i|n_{root})$ denotes the accumulated execution cost resulting from following the trajectory form tree root n_{root} to n_i, and $\hat{c}(x_{goal}|n_i)$ is the estimated cost from n_i to the goal state x_{goal} (cost-to-go). ω_1, $\omega_2 \in [0,1]$ are weight factors. n_i.parent represents the index for the parent node of n_i.

Every two neighbor nodes (n_i, n_j) in the tree is connected by an edge $e(n_i, n_j)$, which represents the closed-loop prediction trajectory $\{x_i \ldots x_j\}$ and control policy to steer the vehicle from x_i to x_j.

5.2. Tree Expansion

The CL-RBT motion planning strategy is built upon the standard closed-loop RRT (CL-RRT) originally proposed by [9] and later extended in [17,18]. Similar to the CL-RRT, our CL-RBT algorithm can be split into two primary processes: a tree expansion process that explores the environment by growing the tree and a path selection and execution loop that selects and executes the optimal portion of the tree. Details of the tree expansion process are described in Algorithm 1.

Unlike the standard RRT that generates candidate control inputs randomly or from a look-up table, the tree expansion process of the CL-RBT predicts the feasible trajectory by incorporating a closed-loop system model consisting of the trajectory-following controller and the vehicle dynamics model (Figure 1). Once the n_{near} is determined, the planner generates a reference $\hat{r}$ input (Line 7) using n_{near} and x_{samp}; then $\hat{r}$ is sent to the closed-loop system consisting of the controller and the vehicle dynamics model: the controller generates the control input $\hat{u}$ (Line 8), which is then sent to the vehicle dynamics to predict the state output $\hat{x}$ by forward simulation (Line 9). The predicted trajectory is then checked against environmental constraints (e.g., collision avoidance) to ensure feasibility.

The CL-RBT algorithm extends the CL-RRT tree expansion by incorporating the prediction of the state (position) uncertainty in growing the tree, using the EKF estimator and the covariance propagation approach (Section 4). First, a node$x_{samp} \in X_{free}$ is generated by random sampling in the free space (Line 1). After that, the prediction of state estimation uncertainty is incorporated as heuristic information in the nearest-node selection: The nearest node n_{near} to x_{samp} by a certain metric is determined from the current tree using the following hybrid heuristic information:

$$i^* = \arg\min_i \left(\lambda_1 c(n_i \mid n_{root}) + \lambda_2 \hat{c}(x_{samp} \mid n_i) + \lambda_3 \frac{x_{samp}.\hat{J}_{samp} - n_i.J_i}{n_i.J} \right) \tag{17}$$

where $\lambda_1, \lambda_2, \lambda_3$ are weighting factors for each item. The above heuristic information consists of three components: the exploration heuristic, the optimization heuristic and the uncertainty heuristic.

Algorithm 1. Closed-loop random belief tree: tree expansion.

1	Take a sample x in the reference space
2	Find the nearest node set $N_{near} = \{n_{near(i)}\}$, $(i \geq 1)$ of x_{samp} from tree T, using the hybrid heuristic consisting of the path cost and uncertainty qualification
3	**for** each node n_{near} in the nearest node set N_{near}
4	$k \leftarrow 0$
5	$\hat{x}(t+k) \leftarrow$ final state of N_{near}
6	**while** $\hat{x}(t+k) \in \chi_{free}(t+k)$ and $\hat{x}(t+k)$ has not reached x_{samp} **do**
7	Generate reference input $\hat{r}(t+k)$ from n_{near} to x_{samp}
8	Generate control input $\hat{u}(t+k)$ from feedback control law
9	Simulate $\hat{x}(t+k+1)$ from state propagation model
10	$k \leftarrow k+1$
11	**end while**
12	Generate the predicted trajectory $\bar{x}(t)$, $t \in [t_0, t_1]$
13	Split the predicted trajectory $\bar{x}(t)$ and add intermediate nodes n_i

if $\bar{x}(t) \in X^{free}$, $\forall t \in [t_0, t_1]$ **then**

Add x_{sample} and all intermediate nodes to T_r, **break**

else if all intermediate nodes are feasible

Add intermediate nodes to T_r, **break**

end if

end for

for each newly added feasible node n **do**

Update path execution cost estimates for n (**Algorithm 2**)

simulate the observations $\bar{z}(t)$ and transfer function along the feasible trajectory between n and n.parent

Update the matrix Ψ_n and posterior covariance Σ_n and uncertainty cost $J(\Sigma)$ of n

Update the total cost of n

end for

The exploration heuristic $\hat{c}(x_{samp}|n_i)$ denotes the estimated cost by connecting n_i to the sample x_{samp}, which enables the path planning algorithm to focus on adding new nodes to the tree and quickly exploring the environment.

The optimization heuristic $c(n_i|n_{root})$ denotes the accumulated cost of the path from the root node n_{root} to the candidate node n_i. The purpose of using this optimization heuristic is to bias the tree growth towards paths that reduce the overall accumulated cost.

In order to incorporate the state estimation uncertainty factor into the path planning progress, the CL-RBT adds the uncertainty heuristic in the nearest node selection of the tree expansion. This uncertainty heuristic enables the motion planner to select the node that leads to the lowest posterior state estimation uncertainty if it is connected to x_{samp}, using the predicted state covariance based on the simulated closed-loop state output and observations along the trajectory between candidate nodes and x_{samp}. This is realized by qualifying the posterior uncertainty along the path from n_i and x_{samp} using the linear covariance propagation described in Section 7: given the covariance Σ_i and the associated factored matrix Ψ_i of n_i, the observations $\mathbf{z}_{i:samp}$ and state output $\hat{\mathbf{x}}_{i:samp}$ are predicted first based on the closed-loop system model. After that, the matrix set $S_{i:samp}$, $G_{i:samp}$, $N_{i:samp}$ and the transfer function $\varsigma_{i:samp}$ along the path are approximated. Therefore, the posterior covariance $\hat{\Sigma}_{samp}$ of x_{samp} resulting from moving the MAV along the path between n_i and x_{samp} can be propagated by:

$$\hat{\Psi}_{samp} = \varsigma_{i:samp}\Psi_i = \begin{bmatrix} 0 & I \\ I & N \end{bmatrix}_{i:samp} \begin{bmatrix} 0 & G^{-T} \\ G & SG^{-T} \end{bmatrix}_{i:samp} \begin{bmatrix} \Lambda_i \\ \Pi_i \end{bmatrix} = \begin{bmatrix} \hat{\Lambda}_{samp} \\ \hat{\Pi}_{samp} \end{bmatrix}$$

$$\hat{\Sigma}_{samp} = \hat{\Lambda}_{samp} \hat{\Pi}_{samp}^{-1}$$

hence, the resulting uncertainty cost of x_{samp} can be given by x_{samp}.

$$.\hat{J}_{samp} = tr(\hat{\Sigma}_{samp}) .$$

The tree expansion process may yield one or more candidate feasible nodes and corresponding trajectories. For each candidate node *n* and trajectory, the planner calculates the path cost (Line 21), simulates the observation along the trajectory (Line 22) and, then, further computes the posterior covariance at *n*, as well as the uncertainty qualification of *n* (Line 23). After updating the total cost, the algorithm finally adds the feasible nodes to the current tree *T* (Line 24).

5.3. Path Cost Evaluation

As aforementioned in Algorithm 1, once a feasible node is identified and added to the tree, the CL-RRT algorithm evaluates and updates its associated cost (Lines 21–24). In order to address the multiple requirements of operating in complex, GPS-denied environments, the CL-RBT adopts multiple cost metrics that incorporate various factors into these tasks. In CL-RBT, the cost of each node can be divided into two primary categories: the path execution cost and uncertainty cost.

The path execution cost denotes the cost of moving the MAV along the trajectory from the root node n_{root} to the goal region via the specific node n_i, and it can be given as the following form:

$$C_p(n_i) = \int_{t_0}^{t_i} c[x(t), u(t)]dt$$

where $[t_0, t_i]$ is the time interval of the corresponding path (node sequence) $\{n_{root}...n_i\}$, and $c[x(t), u(t)]$ is the cost metric of the node trajectory (*i.e.*, Euclidian distance, Dubins distance, *etc.*). In the CL-RBT, the path execution cost consists of two primary components: the accumulated path execution cost and the cost-to-go:

$$C_p(n_i) = C(n_i \mid n_{root}) + C_{CostToGo}$$

where $C(n_i|n_{root})$ represents the accumulated execution cost resulting from following the trajectory from n_{root} to n_i. There are two types of cost estimates of the cost-to-go for each node: a lower bound C_{LB} and an upper bound C_{UB}. The lower bound cost-to-go can be given as the distance metric between the MAV's state at the node and the goal state:

$$C_{LB}(n_i) = C(x_{goal} \mid n_i)$$

The update of the upper bound cost-to-go can be described as follows: each time a new feasible node n_i is added to the tree, the

CL-RBT algorithm attempts to connect n_i to the goal by predicting the closed-loop trajectory between n_i and x_{goal}. If the trajectory is feasible, the upper bound cost-to-go of n_i can be given as the cost associated with this trajectory, otherwise the upper bound cost-to-go of n_i is set to ∞:

$$C_{UB} = \begin{cases} \int_{n_j}^{x_{goal}} c(x(t), u(t)) & : \text{ feasible trajectory between } n_j, x_{goal} \text{ exists} \\ \infty & : \text{ no feasible trajectory between } n_j, x_{goal} \end{cases}$$

After that, the CL-RBT propagates the paths back towards the root node to update the upper bound cost-to-go of n_i's affected ancestor nodes: the old upper bound cost-to-go of the parent node is compared with the cost following the newly generated feasible trajectory. If the latter is smaller, the upper bound cost-to-go of the parent node is updated; otherwise, the propagation procedure stops, since there exists a sub-path of the parent with a lower cost-to-go. This update procedure repeats until the current root node is reached. The details of the cost update procedure are shown in Algorithm 2.

Once a feasible trajectory to the goal is identified, the upper bound cost-to-go is used as the estimated cost-to-go of n_i; otherwise, the lower bound cost-to-go is used. Therefore, the overall path execution cost of node n_i can be given as:

$$C(n_i) = C(n_i \mid n_{root}) + C_{\text{CostToGo}} = C(n_i \mid n_{root}) + \begin{cases} C_{UB} & : \text{TrajectoryToGoal} = \text{True} \\ C_{LB} & : \text{TrajectoryToGoal} = \text{False} \end{cases} \tag{18}$$

Algorithm 2. Closed-loop random belief tree: cost update.

$n_i.C_{LB} \leftarrow c(x_{goal} \mid n_i)$

Compute the accumulated cost of the trajectory between n_{root} and n_i : $C(n_i \mid n_{root})$

Generate the trajectory from n_i to goal state $\bar{x}(x_{goal} \mid n_i)$, using the closed-loop system model

Check the feasibility of the closed-loop trajectory $\bar{x}(x_{goal} \mid n_i)$

if $\bar{x}(x_{goal} \mid n_i)$ is feasible **then**

TrajctoryToGoal $\leftarrow$ **True**

$n_i.C_{UB} \leftarrow C(\bar{x}(x_{goal} \mid n_i))$ (the path cost of $\bar{x}(x_{goal} \mid n_i)$)

$n_k \leftarrow n_i$

while $n_k \neq n_{root}$ **and** n_k.parent.$C_{UB} > n_k.C_{UB} + C(n_k.\text{parent} \mid n_k)$

n_k.parent.$C_{UB} \leftarrow n_k.C_{UB} + C(n_k.\text{parent} \mid n_k)$

$n_k \leftarrow n_k$.parent

end while

$n_i.C_{CostToGo} \leftarrow n_i.C_{UB}$

else

TrajctoryToGoal $\leftarrow$ **False**

$n_i.C_{UB} \leftarrow \infty$

$n_i.C_{CostToGo} \leftarrow n.C_{LB}$

end if

$n_i.C \leftarrow C(n_i \mid n_{root}) + n_i.C_{CostToGo}$

return TrajctoryToGoal

(Note that the cost update procedure also identifies feasible paths that connect the goal to the current tree. This enables the CL-RRT to quickly find paths that reach the goal. As the tree grows, more paths to the goal can be found).

The CL-RBT incorporates the qualification of state estimation uncertainty in the cost function of nodes, such that the generated paths ensure accurate state estimation. As discussed previously, the state estimation uncertainty is evaluated using the uncertainty cost, e.g., the trace of the predicted covariance: $J(n_i) = tr(\Sigma_{ni})$. As shown in Algorithm 1, once a feasible node is added to the tree and its path execution cost is updated, the CL-RBT predicts its posterior state covariance through the update steps described in Section 4 and computes its uncertainty cost (Algorithm 1, Lines 22, 23). Taking the path execution cost and uncertainty cost into consideration, in CL-RBT, the following multi-objective cost function is used:

$$C(n_i) = \zeta_1 C(n_i \mid n_{root}) + \zeta_2 C_{\text{CostToGo}} + \zeta_3 J(n_i.\Sigma) \tag{19}$$

where C_{CostToGo} denotes the estimated cost-to-go from n_i to the goal state, $C(n_i|n_{root})$ is accumulated along the path from root to n_i, and the first two items correspond to the path-execution cost. $J(n_i.\Sigma)$ is the cost associated with the uncertainty of node n_i, which represents the state uncertainty resulting from following path $\{n_{root}...n_i\}$. ζ_1, ζ_2, ζ_3 are weight factors for adjusting the relative importance of finding a short duration path and reducing state estimation uncertainty.

Since there is a unique path from the root to each of the nodes in the tree, the cost of a node can be used to represent the cost of the associated trajectory. The total cost of the node is used to identify the current best trajectory for execution in the selection and execution procedure, which is presented in the next section.

5.4. Path Selection and Execution

In order to account for the changes in time-varying operating environments, the motion planner operates by periodically selecting and executing the best portion of the current trajectory while continuously expanding the tree during the execution process. The path selection and execution process is presented in Algorithm 3.

Algorithm 3. Closed-loop random belief tree: execution loop.

1	$t \leftarrow 0$
2	Initialize tree $\mathcal{T}$ from node at $x_{initial}$
3	**while** $x(t) \notin X_{goal}$ **do**
4	Update the current actual state $x(t_0)$ and environment information
5	Predict state $x(t)$ to $\bar{x}(t_0 + \Delta t)$ and simulate observations
6	**while** time remaining $< \Delta t$ **do**
7	expand the tree using Algorithm 1
8	**end while**
9	Select the current best feasible path (node sequence) $p^* = \{n_0 ... n_k\}$ according to the multiple-objective cost function given as Equation (19)
10	**if** no best feasible path exists **then**
11	Send brake command to the controller and **goto** 3
12	**else**
13	Send p^* to controller to execute
14	**end if**
15	$t \leftarrow t + \Delta t$
16	**end while**

Denote the period for the path selection and execution as Δt. At the beginning of each iteration period, the algorithm updates the current actual state of the MAV, as well as the environment information. The purpose of this step is to ensure the MAV's situational awareness (Line 4). After that, the state is first propagated to the end of the period, resulting in $\bar{x}(t_0 + \Delta t)$ (Line 4). Then, the current root of the tree is set to the most current node that is followed by the propagated state, and all other children nodes of the previous root are removed. This is because the MAV is considered to have passed the previous root, and the paths of its children nodes will never be executed. During each of the iterations, the planner identifies the best portion of the trajectory in terms of the cost metric. Since each node n_i in the tree is associated with a single path, the trajectory can be uniquely specified using the sequence consisting of all of the nodes from root to n_i:$\{n_{root}...n_i\}$. In CL-RBT, the cost function given in Equation (19) is used for selecting the current best paths. This multi-objective cost metric enables the motion planner to select paths that achieve a balance between finding low-cost paths to the goal and minimizing localization uncertainty. At the end of each path selection iteration step, the selected feasible trajectory is sent to the controller to be executed, while the tree keeps expanding (Algorithm 1) during the remaining time of period Δt. This mechanism guarantees that the path planning algorithm can provide a feasible path for execution in real time and account for uncertainty in a dynamic environment.

The overall diagram of the CL-RBT algorithm is depicted in Figure 2.

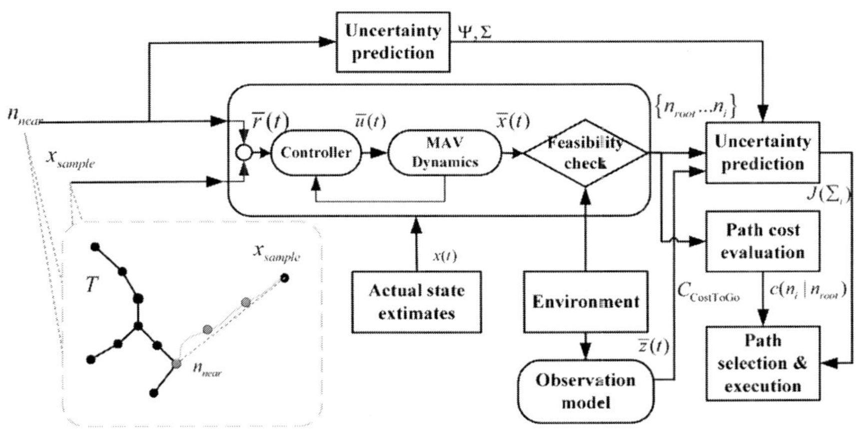

Figure 2. Diagram of the closed-loop random belief tree motion planning framework. MAV, micro-aerial vehicle.

6. INDOOR FLIGHT SIMULATION RESULTS BASED ON A SIMULATED LASER SCANNER-EQUIPPED MAV

In order to validate the effectiveness of the CL-RBT motion planner, we have conducted a number of experiments in various simulated scenarios that are derived from real-world environments. The first scenario consists of a quadrotor MAV navigating in a 3D wide-open indoor environment, with all structures that can be detected by laser rangefinders concentrated along the peripheral walls. The first scenario is used to demonstrate the CL-RBT's ability to generate the path that ensures the MAV's localization confidence. The subsequent scenarios involve more complex extensions, including 3D unstructured environments (the second scenario), and the purpose of these scenarios is to validate the effectiveness of the CL-RBT algorithm in generating paths that satisfy multiple constraints imposed by dynamic, unstructured environments and non-linear MAV dynamics, while achieving a trade-off between minimizing path cost and reducing localization uncertainty in GPS-denied environments.

For all of these experiments, we assume that the dynamics model of the quadrotor is non-linear and that the environment is without access to GPS signals. The quadrotor must start from an initial spot and traverse through the obstacles to a goal location, relying on a path-following control law and the state estimates provided by an EKF estimator. The exteroceptive sensor equipped on the quadrotor is assumed to be a simulated laser rangefinder with a maximum sensing range of 2 m and a 240° field-of-view. For each of these environments, a 3D model of the environment is first generated using an RGB-D-based environmental modeling approach from our previous work [26]. The 3D point-cloud model is then transformed into a polyhedron-based model, which is used for path planning. The CL-RBT is implemented in MATLAB® and all simulations are performed in real time on an Intel 3.2 GHz platform with a 4 G RAM. The CL-RBT algorithm selects and executes the best parts of the paths every Δt seconds.

6.1. Quadrotor Dynamics Model

The quadrotor model that is used in the simulation is depicted in Figure 3. For the simulation experiments, the x-y-z body-fixed coordinates of the quadrotor is derived using the square configuration, and the kinematic and dynamic model of the quadrotor is developed as rigid-body dynamics influenced by the gravity and thrust of the rotors, assuming near-hover aerodynamics and neglecting the effects caused by the translational velocity. The pure pursuit reference law [27] is applied to steer the reference trajectory, and a PD control law is used to control the quadrotor tracking of the reference. This closed-loop system model consisting of the dynamics model and the control law is used for both the trajectory prediction of the CL-RBT planner and the execution of the resultant trajectory.

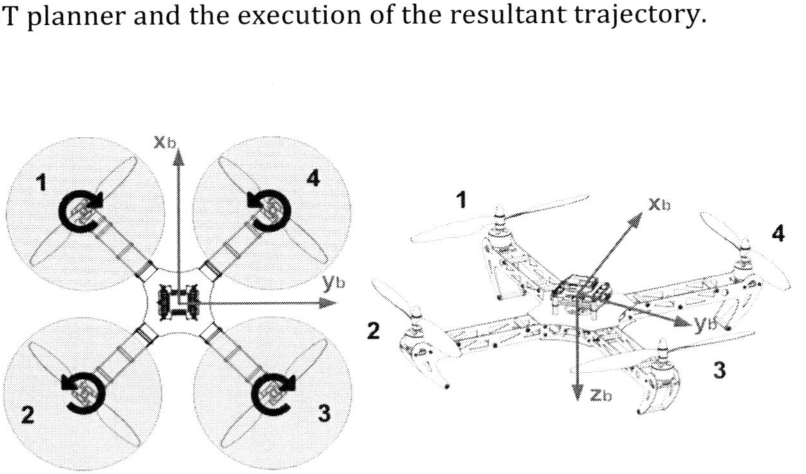

Figure 3. Structure and configuration of the quadrotor MAV.

The configuration of the quadrotor utilized in this paper is illustrated in Figure 3. Two coordinate systems are considered in our scheme: the navigation frame (Earth-fixed frame) ($\mathbf{x}_E, \mathbf{y}_E, \mathbf{z}_E$) and the body-fixed frame ($\mathbf{x}_b, \mathbf{y}_b, \mathbf{z}_b$). The translational and rotational motions of the quadrotors are achieved through the forces and moments caused by varying the angular rates of the four propellers. Assuming that the quadrotor structure is rigid and symmetrical and the center of mass coincides with the origin of the body-fixed frame, the equations of motion of the quadrotor can be derived using Newton's law and Euler equations [27]. Define φ, θ, ψ as the Euler angles denoting the roll, pitch and yaw attitude, respectively, the rotational dynamics can be described as:

$$
\begin{aligned}
\ddot{\varphi} &= \dot{\theta}\dot{\psi}\left(\frac{I_{yy}-I_{zz}}{I_{xx}}\right)+\frac{l}{I_{xx}}b(\omega_3^2+\omega_4^2-\omega_1^2-\omega_2^2)-\frac{K_{drx}}{I_{xx}}\dot{\varphi}+\frac{J_r}{I_{xx}}\dot{\theta}(\omega_2+\omega_4-\omega_1-\omega_3) \\
\ddot{\theta} &= \dot{\varphi}\dot{\psi}\left(\frac{I_{zz}-I_{xx}}{I_{yy}}\right)+\frac{l}{I_{yy}}b(\omega_2^2+\omega_3^2-\omega_1^2-\omega_4^2)-\frac{K_{dry}}{I_{yy}}\dot{\theta}-\frac{J_r}{I_{yy}}\dot{\varphi}(\omega_2+\omega_4-\omega_1-\omega_3) \\
\ddot{\psi} &= \dot{\varphi}\dot{\theta}\left(\frac{I_{xx}-I_{yy}}{I_{zz}}\right)+\frac{1}{I_{zz}}d(\omega_1^2+\omega_3^2-\omega_4^2-\omega_2^2)-\frac{K_{drz}}{I_{zz}}\dot{\psi}
\end{aligned}
\tag{20}
$$

where I_{xx}, I_{yy}, I_{zz} are the moments of inertial around the three axis, J_r is the rotor inertia and l denotes the length of the moment arm. ω_1, ω_2, ω_3, ω_4 are the four propellers' rotation speeds. b and d denote the rotor thrust coefficient and rotor drag coefficient, respectively. K_{drx}, K_{dry}, K_{drz} represent the rotational drag coefficients of the quadrotor body.

Define (x, y, z) as the three-dimensional position with respect to the navigation frame; the translational dynamics of the quadrotor can be given as:

$$\begin{aligned}
\ddot{x} &= (\cos\varphi\cos\psi\sin\theta + \sin\varphi\sin\psi)\frac{1}{m}\sum_{i=1}^{4} T_i - \frac{K_{drx}}{m}\dot{x} \\
\ddot{y} &= (\sin\psi\sin\theta\cos\varphi - \sin\varphi\cos\psi)\frac{1}{m}\sum_{i=1}^{4} T_i - \frac{K_{dry}}{m}\dot{y} \qquad (21) \\
\ddot{z} &= \cos\theta\cos\varphi\frac{1}{m}\sum_{i=1}^{4} T_i - \frac{K_{drz}}{m}\dot{z} - g
\end{aligned}$$

where T_i denotes the thrust generated by propeller i and Ti=bω2i, m is the total mass of the quadrotor and g represents the gravitational acceleration. K_{drx}, K_{dry}, K_{drz} are the translational drag coefficients of the quadrotor body.

6.2. Control Scheme Design of the MAV Model

6.2.1. Cascaded Control Scheme

Under the assumption of near-hovering velocities, the drag terms caused by the translational and rotational motions inEquations (20) and (21) can be neglected, and the system dynamics model can be described in the following state-space form ($\dot{\mathbf{x}} = f(\mathbf{x},\mathbf{u})$):

$$\dot{\mathbf{x}} = \begin{bmatrix} \dot{\varphi} \\ \dot{\theta} \\ \dot{\psi} \\ \dot{\theta}\dot{\psi}\left(\dfrac{I_{yy}-I_{zz}}{I_{xx}}\right)+\dfrac{l}{I_{xx}}bu_2 \\ \dot{\varphi}\dot{\psi}\left(\dfrac{I_{zz}-I_{xx}}{I_{yy}}\right)++\dfrac{l}{I_{yy}}bu_3 \\ \dot{\varphi}\dot{\theta}\left(\dfrac{I_{xx}-I_{yy}}{I_{zz}}\right)+\dfrac{1}{I_{zz}}du_4 \\ \dot{x} \\ \dot{y} \\ \dot{z} \\ (\cos\varphi\cos\psi\sin\theta+\sin\varphi\sin\psi)\dfrac{1}{m}u_1 \\ (\sin\psi\sin\theta\cos\varphi-\sin\varphi\cos\psi)\dfrac{1}{m}u_1 \\ \cos\theta\cos\varphi\dfrac{1}{m}u_1-g \end{bmatrix} \tag{22}$$

where $\mathbf{x} = [\varphi\ \theta\ \psi\ \dot{\varphi}\ \dot{\theta}\ \dot{\psi}\ x\ y\ z\ \dot{x}\ \dot{y}\ \dot{z}]^T$ denotes the state vector, and $\mathbf{u} = [u_1\ u_2\ u_3\ u_4]^T$ represents the control input:

$$\mathbf{u} = \begin{bmatrix} b(\omega_1^2+\omega_2^2+\omega_3^2+\omega_4^2) \\ b(\omega_3^2+\omega_4^2-\omega_1^2-\omega_2^2) \\ b(\omega_2^2+\omega_3^2-\omega_1^2-\omega_4^2) \\ d(\omega_1^2+\omega_3^2-\omega_4^2-\omega_2^2) \end{bmatrix} \tag{23}$$

The physical meaning of the control input is related to the forces generated by the four propellers: u_1 represents the total thrust of the propellers, while u_2, u_3 and u_4 are the differences of the propeller pairs.

As can be seen from Equation (20), the rotational dynamics (φ, θ, ψ) are independent of the translational dynamics (x, y, z), leading to decoupled system dynamics. In addition, the altitude dynamics (z) can also be decoupled from the planar translational dynamics (x, y) by assuming that the φ, θ angles are very small. As a result, the quadrotor dynamics inEquation (21) can be decoupled into independent sub-system dynamics, and the cascaded control systems for the attitude, position and altitude can be designed from the decoupled dynamics. Under the assumption of small φ, θ angles and near-hovering motions, we can derive the following linearized attitude dynamics:

$$\begin{cases} \ddot{\varphi} = \dfrac{l}{I_x} b u_2 \\ \ddot{\theta} = \dfrac{l}{I_y} b u_3 \\ \ddot{\psi} = \dfrac{d}{I_z} u_4 \end{cases} \tag{24}$$

The linearized translational motion dynamics can be given as:

$$\begin{cases} \ddot{x} = (\theta \cos\psi + \varphi \sin\psi) \dfrac{1}{m} u_1 \\ \ddot{y} = (\theta \sin\psi - \varphi \cos\psi) \dfrac{1}{m} u_1 \\ \ddot{z} = \dfrac{1}{m} u_1 - g \end{cases} \tag{25}$$

Using the decoupled rotational and translational dynamics in Equations (24) and (25), we can design a cascaded control scheme consisting of the inner control loop and outer control loop, of which the control laws can be designed separately. As illustrated in Figure 4, the outer control loop is the position and altitude controller, which receives the reference path from the CL-RBT planner and generates the reference roll and pitch commands (φ_r, θ_r) to the inner attitude control loop, while the attitude controller produces the desired rotor speed to achieve the reference attitude according to the attitude commands φ_r, θ_r, ψ_r.

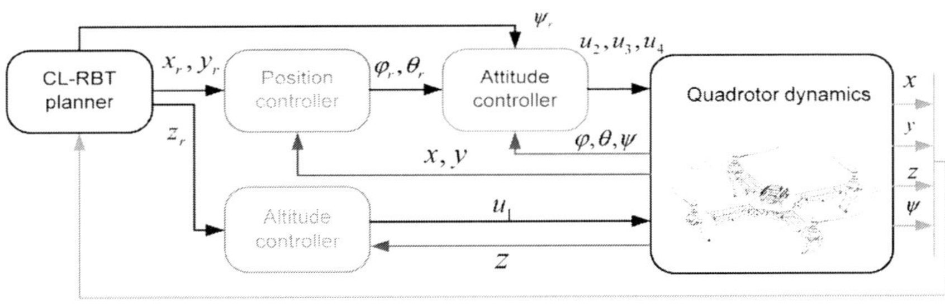

Figure 4. Cascaded control scheme of the quadrotor MAV. CL-RBT, closed-loop random belief trees.

6.2.2. Controller Design

Based on the attitude and translational motion dynamics described in Equations (24) and (25), the proportional-integral-derivative (PID) control scheme is utilized to design the attitude and position controller for the quadrotor MAV. The first step is to consider the inner loop controller, which functions as the core of the control scheme. Using the attitude dynamics shown in Equation (24), the attitude controller can be designed as follows:

$$
\begin{aligned}
u_2 &= K_P^{\varphi,\theta}(\varphi_r-\varphi)+K_I^{\varphi,\theta}\int(\varphi_r-\varphi)dt+K_D^{\varphi,\theta}(\dot{\varphi}_r-\dot{\varphi})\\
u_3 &= K_P^{\varphi,\theta}(\theta_r-\theta)+K_I^{\varphi,\theta}\int(\theta_r-\theta)dt+K_D^{\varphi,\theta}(\dot{\theta}_r-\dot{\theta})\\
u_4 &= K_P^{\psi}(\psi_r-\psi)+K_I^{\psi}\int(\psi_r-\psi)dt+K_D^{\psi}(\dot{\psi}_r-\dot{\psi})
\end{aligned}
\qquad (26)
$$

where $K_P^{\varphi,\theta}, K_I^{\varphi,\theta}, K_D^{\varphi,\theta}, K_P^{\psi}, K_I^{\psi}, K_D^{\psi}$ are the control gains of the PID controller and φ_r, θ_r and ψ_r denote the reference roll, pitch and yaw, respectively. Note that φ_r, θ_r are generated by the outer control loop while ψ_r is directly provided by the CL-RBT planner.

Under the assumption of near-hovering velocities and small rotational angle motions, the control scheme of translational dynamics can be decoupled into the planar (x, y) and altitude controllers that can be designed separately. The altitude controller is designed based on PID control with non-linear compensation:

$$
u_1 = \frac{1}{\cos\theta\cos\varphi}\left(K_P^{z}(z_r-z)+K_I^{z}\int(z_r-z)dt+K_D^{z}(\dot{z}_r-\dot{z})\right) \qquad (27)
$$

where $K_P^{z}, K_I^{z}, K_D^{z}$ are the PID control gains, and z_r is the reference altitude of the path provided by the CL-RBT planner. The planar position controllers take the following form:

$$\theta_r = \cos\psi[K_P^x(x_r - x) + K_I^x \int (x_r - x)dt + K_D^x(\dot{x}_r - \dot{x})]$$
$$+ \sin\psi[K_P^y(y_r - y) + K_I^y \int (y_r - y)dt + K_D^y(\dot{y}_r - \dot{y})]$$
$$\phi_r = \sin\psi[K_P^x(x_r - x) + K_I^x \int (x_r - x)dt + K_D^x(\dot{x}_r - \dot{x})]$$
$$- \cos\psi[K_P^y(y_r - y) + K_I^y \int (y_r - y)dt + K_D^y(\dot{y}_r - \dot{y})]$$

where $K_P^x, K_I^x, K_D^x, K_P^y, K_I^y, K_D^y$ are the PID control gains and x_r, y_r are the reference position provided by the CL-RBT planner. As mentioned before, the outer loop controller outputs θ_r, φ_r to the inner loop controller as the reference pith and roll, which are used in Equation (26) for attitude control.

In our applications, the frequency of the inner loop (attitude controller) is 250 Hz, while the outer loop (translational and altitude controller) runs at 30 Hz. The PID gains of the controllers are tuned through extensive numerical simulations in order to achieve a desirable control performance.

In this paper, we only report simulation results based on the PID control law. The reason for selecting the PID scheme is that it can handle complex dynamics model and it is robust to modeling errors. Note that more complex and advanced control laws can also be adopted in the CL-RBT framework. In addition, other types of MAV dynamics models with associated guidance and control laws can be incorporated to form the closed-loop model in CL-RBT.

6.3. EKF Process Model

In the simulation experiments, the states of the MAV are estimated using an EKF-based estimator. The MAV states to be estimated in the EKF filter include: three-axis position in the global frame $\mathbf{p}$ =

$[x, y, z]^T$, MAV orientation represented using Euler angles $[\varphi, \theta, \psi]^T$ (roll, pitch and yaw), three-axis velocity in the global frame $\mathbf{v} = [v_x, v_y, v_z]^T$, as well as the gyroscope bias $\mathbf{b}_\omega = [b_{\omega x}, b_{\omega y}, b_{\omega z}]^T$ and accelerometer bias $\mathbf{b}_f = [b_{fx}, b_{fy}, b_{fz}]^T$. The estimated state vector can be denoted specifically as follows:

$$\mathbf{x} = [\mathbf{p}, \mathbf{v}, \varphi, \theta, \psi, \mathbf{b}_\omega, \mathbf{b}_f]^T = [x, y, z, v_x, v_y, v_z, \varphi, \theta, \psi, b_{\omega x}, b_{\omega y}, b_{\omega z}, b_{fx}, b_{fy}, b_{fz}]^T \quad (28)$$

As described previously, the MAV is equipped with a laser rangefinder, which is used by scan-matching to provide position and heading measurements (x, y, ψ). The altitude of the MAV is measured by an onboard sonar altimeter, and it is assumed that the altitude measurement is not affected by the attitude of the MAV. In addition to the laser rangefinder, the MAV is assumed to be equipped with an IMU module, which consists of a triaxial gyroscope and triaxial accelerometer. The triaxial gyroscope provides three-axis angular rates $\omega_m = [\omega_{mx}, \omega_{my}, \omega_{mz}]^T$, while the accelerometer measures the three-axis accelerations $f_m = [f_{mx}, f_{my}, f_{mz}]^T$, and the above IMU measurements are all expressed in the MAV's body frame. Using the above definitions, the continuous state-space-based process model $(\mathbf{x} = g(\mathbf{x}, \mathbf{u}))$ describing the IMU dynamics can be given by:

$$\dot{\mathbf{p}} = \mathbf{v} \quad (29)$$

$$\dot{\mathbf{v}} = \tilde{C}_b^n (f_m - \mathbf{b}_f) - \mathbf{g} \quad (30)$$

$$\begin{bmatrix} \dot{\phi} \\ \dot{\theta} \\ \dot{\psi} \end{bmatrix} = \begin{bmatrix} 1 & \sin\phi\tan\theta & \cos\phi\tan\theta \\ 0 & \cos\phi & -\sin\phi \\ 0 & \sin\phi\sec\theta & \cos\phi\sec\theta \end{bmatrix} \begin{bmatrix} \omega_{mx} - b_{\omega x} \\ \omega_{my} - b_{\omega y} \\ \omega_{mz} - b_{\omega z} \end{bmatrix} \quad (31)$$

$$\dot{\mathbf{b}}_{\omega} = 0 \quad (32)$$

$$\dot{\mathbf{b}}_{f} = 0 \quad (33)$$

where $\tilde{C}_b^n$ denotes the transformation matrix from the body frame to the global frame, and $\mathbf{g} = [0, 0, g]^T$ is the local gravity vector in the global frame. Since the x, y position and the heading ψ (yaw) of the MAV are estimated separately using the measurements from the laser scan-matching, ψ can be treated independent of roll φ and pitch θ, and the MAV motion of the$x$-, y-axes can be decoupled from z-axis motion under the assumption of near-hovering velocity and small φ, θ angles. This assumption yields the simplified transformation matrix in Equation (30):

$$\tilde{C}_b^n = \begin{bmatrix} \cos\psi & -\sin\psi & 0 \\ \sin\psi & \cos\psi & 0 \\ 0 & 0 & 1 \end{bmatrix} \quad (34)$$

A similar approach for the simplification of the MAV motion can also be found in [28,29] (Note that although this simplified velocity model may cause significantly large estimation error when the MAV performs aggressive attitude maneuvers or high-speed flight, it is sufficiently accurate for state estimation under the above assumptions in this paper).

To implement the state estimation on a computer system, the above process model is transformed into a discrete-time model. Let Δt be the update period of EKF and assume that Δt is sufficiently small, the process model in Equations (29)–(33)can be discretized as follows:

$$\mathbf{x}_t = \mathbf{x}_{t-1} + \Delta t f(\mathbf{x}_{t-1}, \mathbf{u}_{t-1}) \tag{35}$$

In order to implement the EKF estimator, the above discrete process model needs to be further linearized by calculating the Jacobians $\partial g / \partial \mathbf{x}|_{\mathbf{x}_{t-1}}$ as described in Equation (2). To derive the Jacobians, the calculation of partial derivatives of the process model with respect to the state vector is given as follows:

$$G = \left.\frac{\partial g}{\partial \mathbf{x}}\right|_{\mathbf{x}_{t-1}} = I + \Delta t \left.\frac{\partial f}{\partial \mathbf{x}}\right|_{\mathbf{x}_{t-1}} = \begin{bmatrix} I_{3\times3} & \Delta t I_{3\times3} & 0_{3\times3} & 0_{3\times3} & 0_{3\times3} \\ 0_{3\times3} & I_{3\times3} & \Delta t G_{23} & 0_{3\times3} & \Delta t G_{25} \\ 0_{3\times3} & 0_{3\times3} & I_{3\times3} + \Delta t G_{33} & \Delta t G_{34} & 0_{3\times3} \\ 0_{3\times3} & 0_{3\times3} & 0_{3\times3} & I_{3\times3} & 0_{3\times3} \\ 0_{3\times3} & 0_{3\times3} & 0_{3\times3} & 0_{3\times3} & I_{3\times3} \end{bmatrix} \tag{36}$$

where:

$$G_{23}=\begin{bmatrix}0 & 0 & -\sin\psi(f_{mx}-b_{fx})-\cos\psi(f_{my}-b_{fx})\\ 0 & 0 & \cos\psi(f_{mx}-b_{fx})-\sin\psi(f_{my}-b_{fx})\\ 0 & 0 & 0\end{bmatrix},\ G_{25}=\begin{bmatrix}-\cos\psi & \sin\psi & 0\\ -\sin\psi & -\cos\psi & 0\\ 0 & 0 & -1\end{bmatrix}$$

$$G_{33}=\begin{bmatrix}(\omega_{my}-b_{\omega y})\cos\varphi\tan\theta-(\omega_{mz}-b_{\omega z})\sin\varphi\tan\theta & (\omega_{my}-b_{\omega y})\sin\varphi\sec^2\theta+(\omega_{mz}-b_{\omega z})\cos\varphi\sec^2\theta & 0\\ -(\omega_{my}-b_{\omega y})\sin\varphi-(\omega_{mz}-b_{\omega z})\cos\varphi & 0 & 0\\ (\omega_{my}-b_{\omega y})\cos\varphi\sec\theta-(\omega_{mz}-b_{\omega z})\sin\varphi\sec\theta & (\omega_{my}-b_{\omega y})\sin\varphi\sin\theta\sec^2\theta+(\omega_{mz}-b_{\omega z})\cos\varphi\sin\theta\sec^2\theta & 0\end{bmatrix}$$

$$G_{34}=\begin{bmatrix}-1 & -\sin\varphi\tan\theta & -\cos\varphi\tan\theta\\ 0 & -\cos\varphi & \sin\varphi\\ 0 & -\sin\varphi\sec\theta & -\cos\varphi\sec\theta\end{bmatrix}$$

The process noise w is introduced by the IMU angular rates (ω_m) and accelerations (f_ω), which are included as the input **u**to the process model:

$$\mathbf{u}=\left[\omega_m, f_m,\right]^{\mathrm{T}},\ w=\left[w_\omega, w_f\right]^{\mathrm{T}} \tag{37}$$

where w_ω, w_f are the noise associated with the angular rates and accelerations, respectively, and w_ω, w_f are both white Gaussian noises with zero mean, *i.e.*, $w_\omega \sim N(0, Q_\omega)$, $w_f \sim N(0, Q_f)$. Similarly, the partial derivatives of the process model with respect to the process noise can be calculated by:

$$V_t = \left.\frac{\partial g}{\partial w}\right|_{\mathbf{x}_{t-1},\mathbf{u}_{t-1}} = \Delta t \left.\frac{\partial f}{\partial w}\right|_{\mathbf{x}_{t-1},\mathbf{u}_{t-1}} = \begin{bmatrix} 0_{3\times3} & 0_{3\times3} & \Delta t V_{31} & 0_{3\times3} & 0_{3\times3} \\ 0_{3\times3} & \Delta t V_{22} & 0_{3\times3} & 0_{3\times3} & 0_{3\times3} \end{bmatrix}^{\mathrm{T}} \quad (38)$$

where:

$$V_{22} = \tilde{C}_b^n, V_{31} = \begin{bmatrix} 1 & \sin\varphi\tan\theta & \cos\varphi\tan\theta \\ 0 & \cos\varphi & -\sin\varphi \\ 0 & \sin\varphi\sec\theta & \cos\varphi\sec\theta \end{bmatrix}$$

Due to the different measurement rates of the IMU, the sonar and the scan-matching module, the state and covariance are updated separately at different times, using three groups of measurements. The pitch, roll angle (φ, θ) and the gyroscope bias $\mathbf{b}_\omega$ are updated first using the measurements from the IMU (more specifically, the accelerometer measurements). The altitude z is updated separately using the sonar measurement, since it is uncorrelated with the other measurements. In addition, the update of the MAV position $\mathbf{p}$, heading ψ, the velocity v_x, v_y and the accelerometer bias $\mathbf{b}_f$ are completed when the measurements from laser scan-matching are received. By the above discussions, while the performance of pitch and roll (φ, θ) estimates depends on the IMU, the estimation of x, y position and the heading ψ are primarily affected by the characteristics of the laser rangefinder, as well as the perceived environment information. As a result, in order to evaluate how the path planning strategy affects the uncertainty of state estimation, we extract the laser-related states $\tilde{\mathbf{x}}=[x,y,v_x,v_y,\psi,b_{fx}, b_{fy}]^{\mathrm{T}}$ from the full state vector. By Equation (35), the process update model of $\tilde{\mathbf{x}}$ can be given as:

$$
\begin{aligned}
&x_t = x_{t-1} + v_{x|t-1}\Delta t \\
&y_t = y_{t-1} + v_{y|t-1}\Delta t \\
&\begin{bmatrix} v_x \\ v_y \end{bmatrix}_t = \begin{bmatrix} v_x \\ v_y \end{bmatrix}_{t-1} + \Delta t \begin{bmatrix} \cos\psi & -\sin\psi \\ \sin\psi & \cos\psi \end{bmatrix}_{t-1} \begin{bmatrix} f_{mx} - b_{fx} \\ f_{my} - b_{fy} \end{bmatrix}_{t-1} + \Delta t \begin{bmatrix} \cos\psi & -\sin\psi \\ \sin\psi & \cos\psi \end{bmatrix}_{t-1} \begin{bmatrix} w_{fx} \\ w_{fy} \end{bmatrix} \\
&\psi_t = \psi_{t-1} + \Delta t(\omega_{mz|t-1} - b_{\omega z|t-1}) + \Delta t w_{\omega z} \\
&b_{fx|t} = b_{fx|t-1} \\
&b_{fy|t} = b_{fy|t-1}
\end{aligned}
$$

(39)

By the above process model, we have:

$$\tilde{\mathbf{x}}_t = \tilde{g}(\tilde{\mathbf{x}}_{t-1}, \tilde{\mathbf{u}}_{t-1}, \tilde{w}) \tag{40}$$

where $\tilde{\mathbf{u}} = \left[f_{mx}, f_{my}, \omega_{mz} \right]^{\mathrm{T}}$, $\tilde{w} = \left[w_{fx}, w_{fy}, w_{\omega z} \right]^{\mathrm{T}}$, and the Jacobians of the above model can be obtained directly from Equations (36)and (38):

$$
\tilde{G}_t = \left.\frac{\partial g}{\partial \mathbf{x}}\right|_{\mathbf{x}_{t-1}, \mathbf{u}_{t-1}} = \begin{bmatrix}
1 & 0 & \Delta t & 0 & 0 & 0 & 0 \\
0 & 1 & 0 & \Delta t & 0 & 0 & 0 \\
0 & 0 & 1 & 0 & -\Delta t \sin\psi(f_{mx} - b_{fx}) - \Delta t \cos\psi(f_{my} - b_{fy}) & -\Delta t \cos\psi & \Delta t \sin\psi \\
0 & 0 & 0 & 1 & \Delta t \cos\psi(f_{mx} - b_{fx}) - \Delta t \sin\psi(f_{my} - b_{fy}) & -\Delta t \sin\psi & -\Delta t \cos\psi \\
0 & 0 & 0 & 0 & 1 & 0 & 0 \\
0 & 0 & 0 & 0 & 0 & 1 & 0 \\
0 & 0 & 0 & 0 & 0 & 0 & 1
\end{bmatrix}
$$

(41)

$$\tilde{V}_t = \begin{bmatrix} 0 & 0 & \Delta t\cos\psi & \Delta t\sin\psi & 0 & 0 & 0 \\ 0 & 0 & -\Delta t\sin\psi & \Delta t\cos\psi & 0 & 0 & 0 \\ 0 & 0 & 0 & 0 & \Delta t & 0 & 0 \end{bmatrix}^{\mathrm{T}} \qquad (42)$$

Denote σ_{fx}^2, σ_{fy}^2 and $\sigma_{\omega z}^2$ as the noise covariance of the accelerometer measurements in the *x*- and *y*-axis, as well as the gyroscope measurements in the *z*-axis, respectively. The process noise covariance of the process model in Equation (39) can be given by:

$$Q = diag(\sigma_{fx}^2, \sigma_{fy}^2, \sigma_{\omega z}^2) \qquad (43)$$

Therefore, the matrix S_t in Equation (11) can be calculated by $S_t = \tilde{V}_t Q \tilde{V}_t^T$, and $\tilde{G}_t$ (Equation (41)), $\tilde{V}_t$ (Equation (42)), S_t are used in the linear covariance propagation, as described in Equation (11).

6.4. Sensor Measurement Model and Uncertainty Analysis of Laser Rangefinders

In many actual applications, a MAV performs localization using an onboard laser rangefinder, which obtains measurements through a series of range readings recorded at consecutive angles increments. Therefore, the accuracy of the localization is mainly affected by the range to the objects and the topology of the environments. In our simulation, we assume that the quadrotor is equipped with a laser rangefinder and that the localization is performed by matching the laser scans. In order to incorporate the uncertainty of the laser rangefinder for the covariance propagation in the CL-RBT planner, we used an information matrix-based method [6] to determine the uncertainty of the laser measurements with respect to the topology of the environments.

For a quadrotor using a laser rangefinder, the information matrix *N* in Equation (11) denotes the information provided by laser scans, which can be calculated by combining the information of each range measurement of the environment. Considering an environment geometry shown in Figure 5, a laser scan obtained at time *t* consists of measurements of *n* scan points in the environment, each of which is described by a range reading r_i and a measurement angle θ_i, and the information provided in a laser scan (r_i, θ_i) is in the direction perpendicular to the topology, *i.e.*, the direction of the vector normal (the vector in green, Figure 5) of the environment surface. Therefore, the information matrix can be obtained by projecting the range information onto the vector normal of the surface and summing the projected information of each scan point. The information matrix *N* can be calculated by:

$$\mathbf{N} \triangleq \mathbf{H}^T (\Sigma_r^2)^{-1} \mathbf{H} = \begin{bmatrix} \Sigma_{xx} & \Sigma_{xy} & \Sigma_{x\psi} \\ \Sigma_{xy} & \Sigma_{yy} & \Sigma_{y\psi} \\ \Sigma_{x\psi} & \Sigma_{y\psi} & \Sigma_{\psi\psi} \end{bmatrix} \quad (44)$$

with **H** as the measurement matrix (Equation (45)) and Σ_r^2 as the variance matrix (Equation (46)) denoting the variance of each measurement:

$$\mathbf{H} = \begin{bmatrix} \cos\gamma_1 \cos(\gamma_1 - \theta_1) & \sin\gamma_1 \cos(\gamma_1 - \theta_1) & r_1 \sin(\gamma_1 - \theta_1) \\ \vdots & \ddots & \vdots \\ \cos\gamma_n \cos(\gamma_n - \theta_n) & \cos\gamma_n \cos(\gamma_n - \theta_n) & r_i \sin(\gamma_n - \theta_n) \end{bmatrix} \quad (45)$$

$$\Sigma_r^2 = diag(\sigma_{r_1}^2, \sigma_{r_1}^2, \sigma_{r_1}^2, \ldots \sigma_{r_i}^2, \sigma_{r_i}^2, \sigma_{r_i}^2 \ldots \sigma_{r_n}^2, \sigma_{r_n}^2, \sigma_{r_n}^2) \quad (46)$$

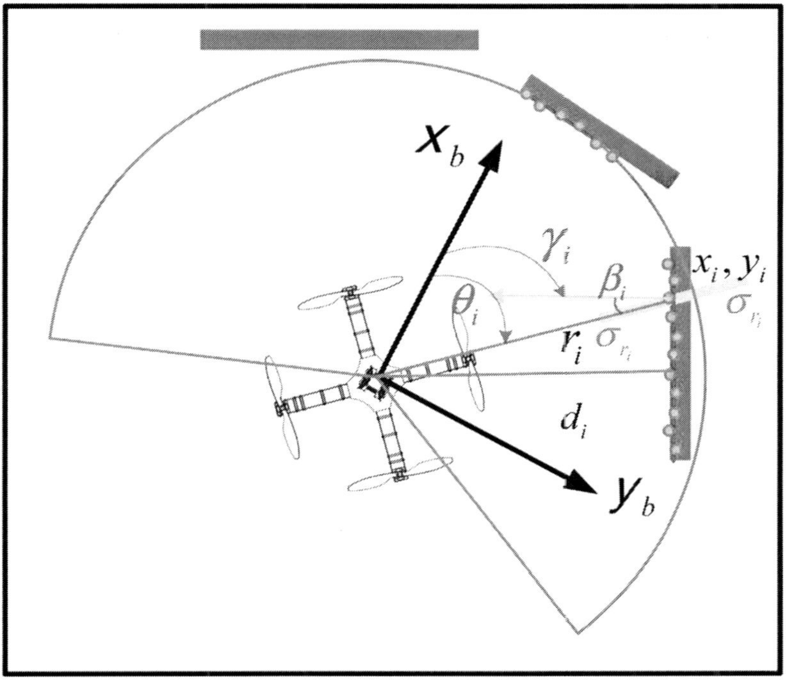

Figure 5. Uncertainty analysis of a laser range measurement. The blue points denote the measurement points in a single laser scan at time *t*. γ is the angle between the body axis and the vector normal of the surface at measurement point (x_i, y_i), and β_i is the angle form the vector normal to the measurement direction. *d* is the distance from the origin to the line segment *i*.

In our experiments, the readings of each measurement point (r_i, θ_i) are obtained first from the simulated laser scan, and then, the line segments of the topology surface are extracted using the measurement points, along with the vector normal of each line segments, as well as the angle between the body axis and the vector normal. From Equations (44)–(46), the information matrix N can be calculated with the summation terms that are given as follows:

$$\Sigma_{xx} = \sum_i^m \frac{\cos^2 \gamma_i \cos^2(\gamma_i - \theta_i)}{\sigma_{r_i}^2}, \Sigma_{yy} = \sum_i^m \frac{\sin^2 \gamma_i \cos^2(\gamma_i - \theta_i)}{\sigma_{r_i}^2}$$

$$\Sigma_{\psi\psi} = \sum_i^m \frac{r_i^2 \sin^2(\gamma_i - \theta_i)}{\sigma_{r_i}^2}, \Sigma_{xy} = \sum_i^m \frac{\cos \gamma_i \sin \gamma_i \cos^2(\gamma_i - \theta_i)}{\sigma_{r_i}^2}$$

$$\Sigma_{x\psi} = \sum_i^m \frac{r_i \cos \gamma_i \sin(\gamma_i - \theta_i) \cos(\gamma_i - \theta_i)}{\sigma_{r_i}^2}, \Sigma_{y\psi} = \sum_i^m \frac{r_i \sin \gamma_i \sin(\gamma_i - \theta_i) \cos(\gamma_i - \theta_i)}{\sigma_{r_i}^2}$$

The above analysis provides a statistical approach to incorporating the variance of the laser scan data into the uncertainty of the pose estimate. Therefore, we can use Equations (44)–(46) to calculate the information matrix N, which is used in the linear covariance propagation as described in Equation (11).

Figure 6a–e show the uncertainty ellipses related to the position estimation uncertainties along a predefined path calculated using the information matrix equations proposed in this section, as well as the covariance propagation approaches in Section 4. In the experiments, an 18 × 9 m 3D environment model is derived from a cluttered real-world laboratory, and the MAV is assumed to be equipped with a simulated laser rangefinder that is able to generate range and bearing measurements of the environment. In order to highlight the influences of the environment, the laser sensor model is assumed to have a limited sensing capability with a 2-m range and a 240° field-of-view, which is represented by blue sectors in the figures (Although this sensor model is unrealistic in terms of maximum range, it serves well to illustrate how the environment and sensor measurement affect the localization uncertainty). As the MAV navigates along the path and rotates its orientations, the environment topologies that fall within the field-of-view will generate a series of noisy measurements of the topologies' distance and bearing to the MAV body (the valid measurements are demonstrated by the cyan sectors in the following figures), enabling the MAV to localize itself by a EKF-based approach using these measurements of the environment. The $1-\sigma$ position uncertainty is demonstrated by red ellipses in the figures, and the size and shape of the ellipses indicate the relative scale of the localization uncertainty in both x and y directions.

As can be seen from Figure 6, the area where the onboard sensor is expected to detect more environmental topologies tends to produce position estimates with low uncertainty (Figure 6a,b), while those locations where the sensor only encounters just one obstacle will lead to high localization uncertainty (Figure 6d). In extreme cases where the MAV navigates into wide open areas (Figure 6e), the localization uncertainty increases sharply, since these areas provide almost no information for localization. In addition, the localization uncertainty is also affected by the orientation (yaw) of the MAV relative to the environmental topologies: The MAV is at the same x-y-z coordinates in Figure 6b,c, but the position uncertainty varies due to the different yaw orientations of the sensor (−45° in Figure 6b and 0° in Figure 6c); the yaw angle is positive when the rotation is counterclockwise around the z-axis of the Earth-fixed frame). The environmental topology of the MAV location in Figure 6b,c can be considered as a corridor between two parallel obstacles, and the laser sensor measurements will provide most information (large information matrix N) in the direction perpendicular to the local environment, according to the model given in Equations (44)–(46). As a result, the uncertainty in the direction of the "corridor" (x direction of the environment frame) is much larger when the measurement direction is perpendicular to the "corridor" (Figure 6c).

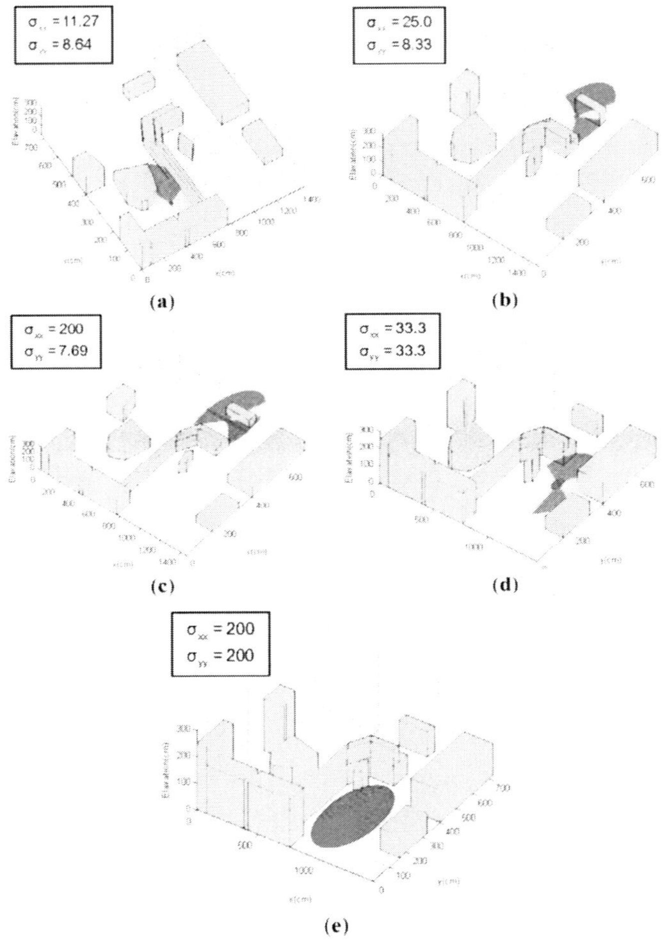

Figure 6. Position estimation uncertainties based on a simulated laser sensor along a predefined path. Cyan sectors denote the laser data with valid environmental measurements. Red ellipses illustrate the 1−σ localization covariance; larger ellipses indicate high uncertainty poses. All covariance are in cm. (**a**) position: (5.0, 2.0, 0.5) m, heading: −45°; (**b**) position: (8.0, 7.5, 0.5) m, heading: −45°; (**c**) position: (8.0, 7.5, 0.5) m, heading: 0°; (**d**) position: (11.0, 4.0, 1.0) m, heading: −30°; (**e**) position: (10.0, 2.5, 1.0) m, heading: 0°.

6.5. Numerical Simulation Results

6.5.1. Scenario 1: Indoor Environment with Open Space

The environment of the first scenario is shown in Figure 7a. As can be seen from Figure 7, this scenario is a 10 × 10-m open indoor environment with all structures concentrated along the peripheral walls, and the center of the environment is an open region that is out of the measurement range of the simulated laser rangefinder. The MAV quadrotor must navigate from the initial position at the bottom-right corner through the hallway to a goal region, which is diagonally opposite of the initial point. Since most of the environment provides rare measurement for localization, it is even more critical for the path planning strategy to find paths that maximize information gain and reduce state uncertainty; hence, the emphasis of this simulation scenario is to test the path planning algorithm's ability to generate paths that ensure the MAV's localization performance throughout the path. For comparison purposes, the conventional CL-RRT algorithm without the incorporation of state uncertainty is also tested using the same MAV model and environment configurations.

In this scenario, the quadrotor MAV begins at $\mathbf{x}_{initial}$ =[2.0 2.0 0.0]^T m. It is intended to navigate towards the goal location at $\mathbf{x}_{goal}$ =[9.0 9.0 1.0]^{T}m, and the weight factors in Equation (19) are set as: ζ_1 = 1, ζ_1 = 1, ζ_3 = 100. The cascaded PID control scheme described in Section 6.2 is applied for both CL-RBT state propagation and reference path following of the MAV. The planning-execution cycle is set to 5 s.

Figure 8 demonstrates an example trajectory generated by a trial of the simulation scenario. The CL-RBT algorithm quickly generates a feasible path that goes through the regions with as much measurements as possible to localize and reaches the goal

(Figure 8a,b). Since a comparatively small amount of samples and nodes are generated during the initial planning, the initially selected paths may still reach the open regions with comparatively high localization uncertainty (Figure 8c–e). However, as more samples and nodes are added, the CL-RBT continuously refines the path in real time, identifying optimal paths that go through measurement-rich regions to ensure localization and reduce the execution cost of the goal (Figure 8f–h).

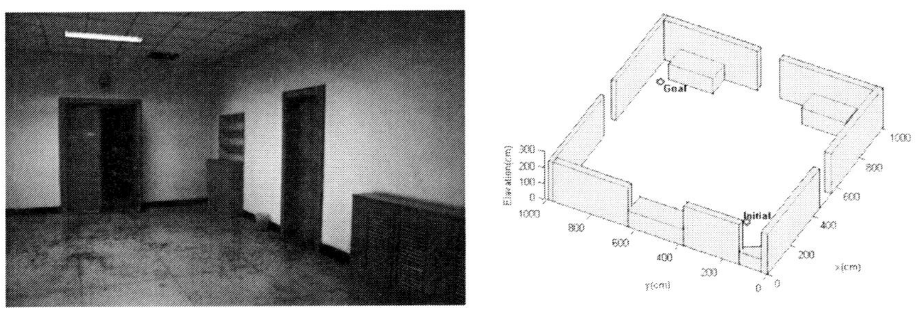

(a) (b)

Figure 7. Indoor environment simulation Scenario 1: (**a**) A hallway of Tsinghua University's main building; (**b**) simulated 3D model of the GPS-denied indoor environment. The red and blue circles indicate the initial and goal location of the MAV, respectively.

In contrast, by expanding the path using the conventional CL-RRT and checking obstacle collision constraints, the resulting paths will move the MAV straight to the goal (Figure 9). Although the conventional CL-RRT may generate shorter paths, ignoring the localization factors will result in high localization uncertainty, since these paths may go through the open regions. In this case, following these paths would likely cause operation failure, since the state estimate becomes unacceptably uncertain, such that the MAV control would have become unstable.

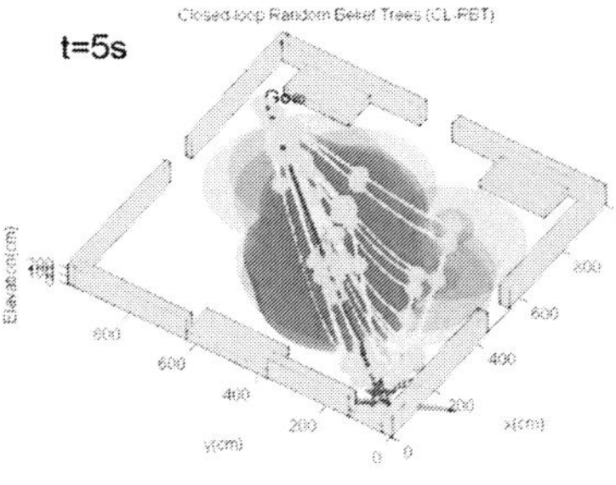
Closed-loop Random Belief Trees (CL-RBT)
t=5s

(a)

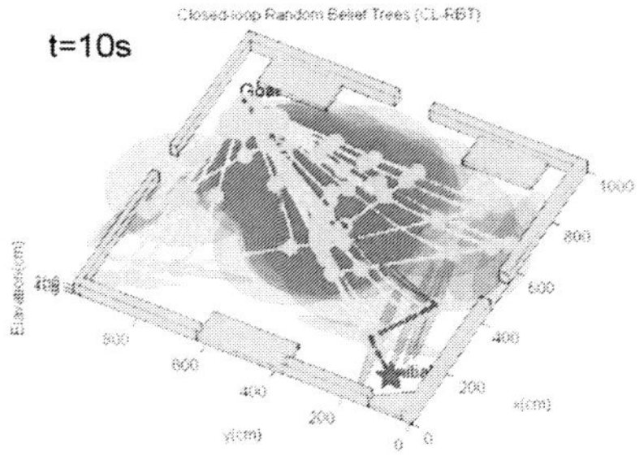
Closed-loop Random Belief Trees (CL-RBT)
t=10s

(b)

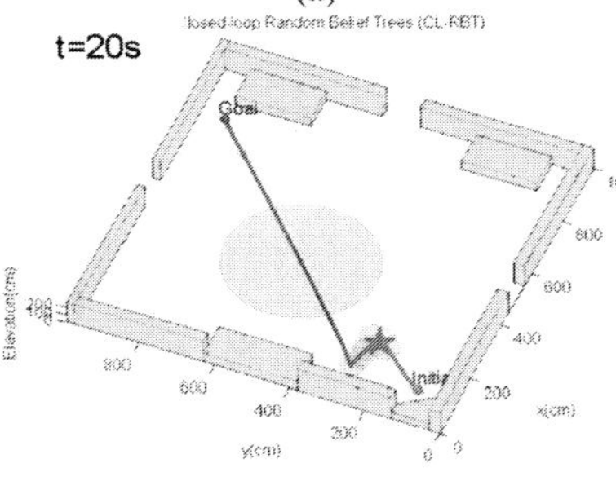
Closed-loop Random Belief Trees (CL-RBT)
t=20s

(c)

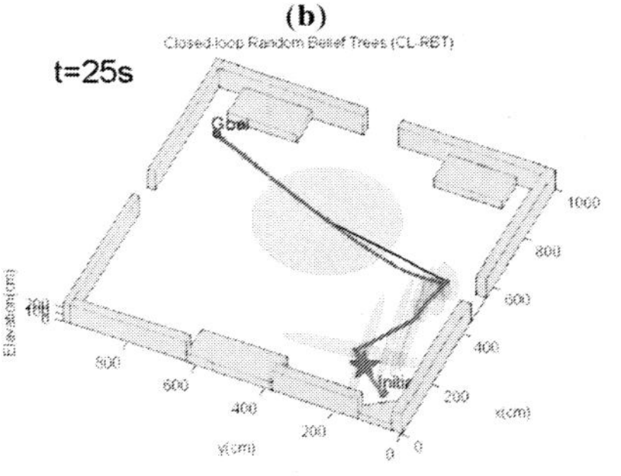
Closed-loop Random Belief Trees (CL-RBT)
t=25s

(d)

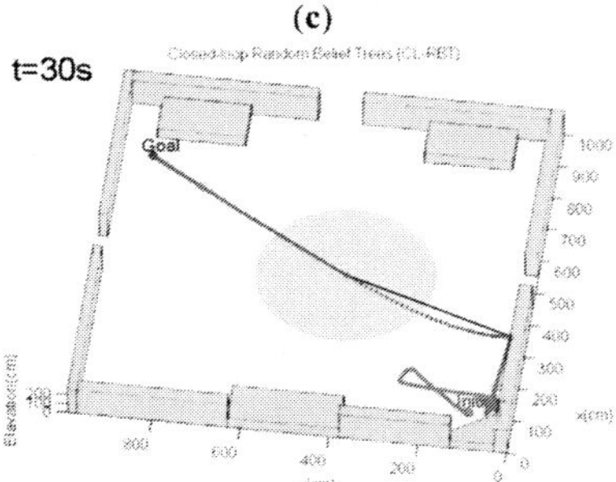
Closed-loop Random Belief Trees (CL-RBT)
t=30s

(e)

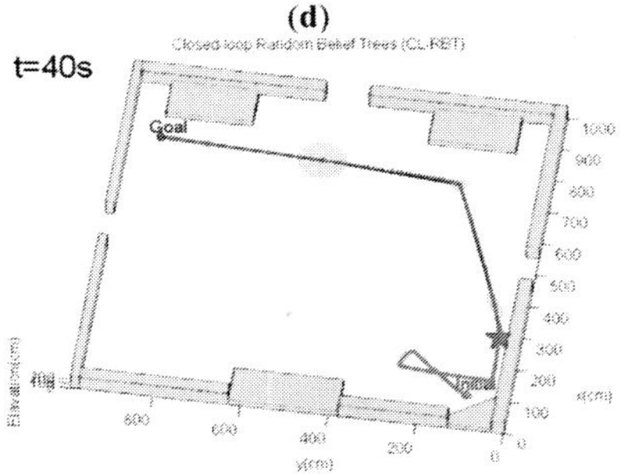
Closed-loop Random Belief Trees (CL-RBT)
t=40s

(f)

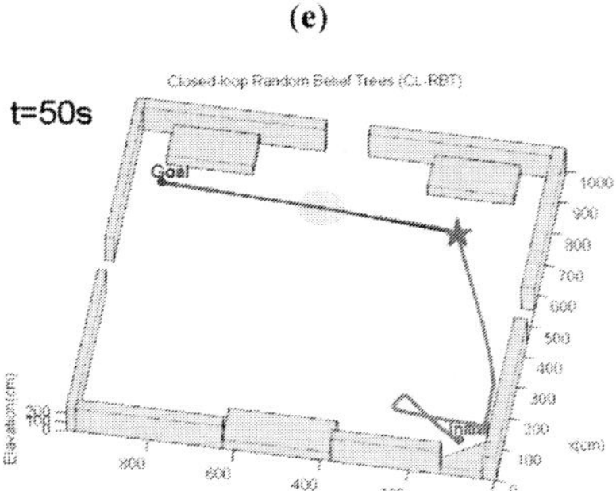
Closed-loop Random Belief Trees (CL-RBT)
t=50s

(g)

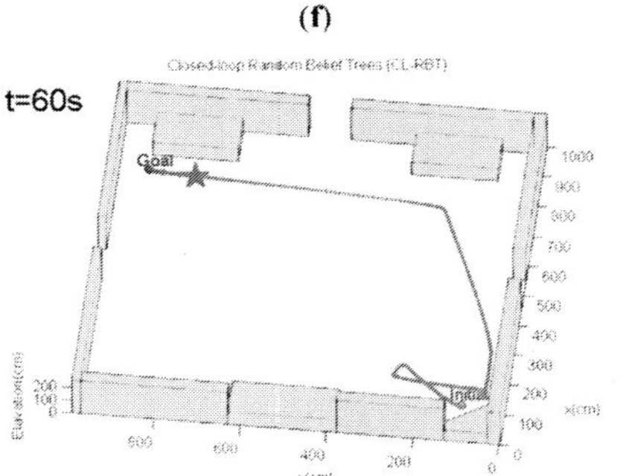
Closed-loop Random Belief Trees (CL-RBT)
t=60s

(h)

Figure 8. Example results of trajectory generated by CL-RBT for a quadrotor MAV navigating in simulation Scenario 1. The pink ellipses represent the covariance of the position estimate, and large ellipses denote MAV states with high uncertainty. The current state (MAV's 3D positions and headings) is denoted with a red pentacle. The current selected path (specified by node sequence, used as the reference to the control system) is marked in black with red dots representing the nodes. The predicted closed-loop trajectory corresponding to the current selected node sequence is emphasized in red, while the actual output trajectory flown by the closed-loop MAV model is marked in pink. The current tree is denoted by green nodes and cyan trajectories, but is set as invisible after (**b**) for clarity. (**a**) t = 5 s; (**b**) t = 10 s; (**c**) t = 20 s; (**d**) t = 25 s; (**e**) t = 30 s; (**f**) t = 40 s; (**g**) t = 50 s; (**h**) t = 60 s.

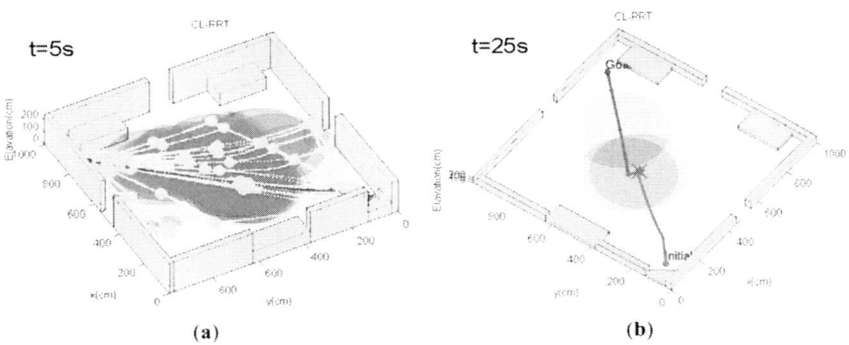

(**a**) (**b**)

Figure 9. Example results of the trajectory generated by the conventional closed-loop rapidly exploring random trees (CL-RRT) for a quadrotor MAV navigating in simulation Scenario 1. The red cross in (**b**) denotes the state where the MAV's state estimation fails, since the position uncertainty has become sufficiently high. (**a**) t = 5 s; (**b**) t = 25 s.

6.5.2. Scenario 2: Cluttered 3D Indoor Environment

This scenario considers a more complex three-dimensional, GPS-denied indoor environment with a number of non-convex obstacles and structures (Figure 10). The environment model is derived from an actual cluttered laboratory, which is the same as the model shown in Figure 6. The objective of the quadrotor MAV is to navigate from the bottom-left initial point to the goal position located on the other side of the laboratory, moving safely between the obstacles. Planning trajectories in this cluttered environment is a challenging task for the path planner. The unstructured environment requires the path planning algorithm to be able to generate a trajectory that avoids the non-convex obstacles to ensure feasibility, while bounding the state estimate uncertainty along the trajectory to allow the MAV to remain well-localized.

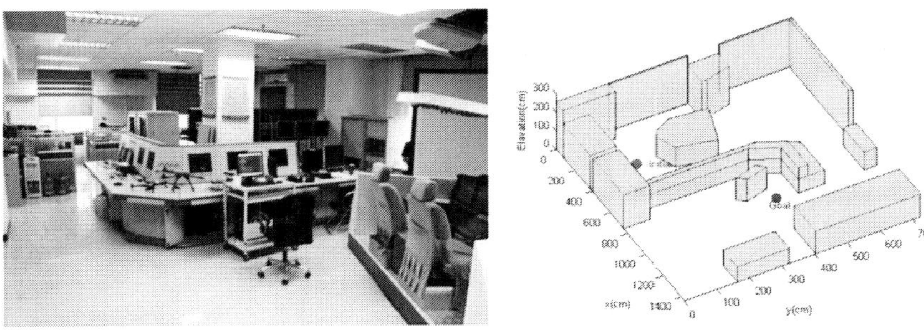

(a) (b)

Figure 10. Indoor environment simulation Scenario 2: (**a**) view of a typical cluttered laboratory; (**b**) simulated 3D model of the unstructured, GPS-denied indoor environment. The red and blue dot indicate the initial and goal location of the MAV, respectively.

For this scenario, the quadrotor MAV begins from one side of the environment at $\mathbf{x}_{initial}$ =[3.2 6.4 0.0]^T, and it is intended to navigate to the goal point behind a cluttered region of obstacles at $\mathbf{x}_{goal}$ =[10.0 4.0 0.5]$^T m$. The same MAV dynamics model, PID control scheme and sensor configuration as in Scenario 1 are also applied in this scenario. However, we select weighting factors in Equation (19) as $\zeta_1 = 1, \zeta_1 = 20, \zeta_3 = 30$, since the emphasis of this scenario is on generating feasible trajectories that quickly reach the goal while reducing localization uncertainty. The planning cycle in this scenario is extended to 7 s. The conventional CL-RRT is also tested using the same scenario and MAV system model for comparison purpose.

An example of the trajectory simulation results from this scenario is depicted in Figure 11. As can be seen from Figure 11, the MAV's initial position is located in an obstacle-rich region. The path planner initially has to move the MAV out of this cluttered region. Unlike the first scenario, the CL-RBT algorithm does not find any trajectories directly to the goal due to the poor coverage of the CL-RBT tree and the cluttered environment in the first few planning cycles (Figure 11a,b). As the CL-RBT tree expands to cover more space after some wandering, the CL-RBT algorithm moves the MAV progressively through the narrow passages and approaches towards the direction of the goal region Figure 11c). After moving out of the narrow passage region, the path planner succeeds in identifying a trajectory to the goal (Figure 11d). As can be seen from the figure, the covariance of the position estimate stays bounded along the trajectory during the entire planning cycles of CL-RBT.

In contrast to CL-RBT, the conventional CL-RRT quickly finds a path that goes over the cluttered obstacle region and directly reaches the goal (Figure 12a). Although the resulting path is shorter than the path generated by CL-RBT, the uncertainty of the position estimate grows rapidly along the CL-RRT trajectory

(Figure 12b); this is because the environment becomes open as the MAV's altitude increases, providing rare information to the onboard sensor for localization.

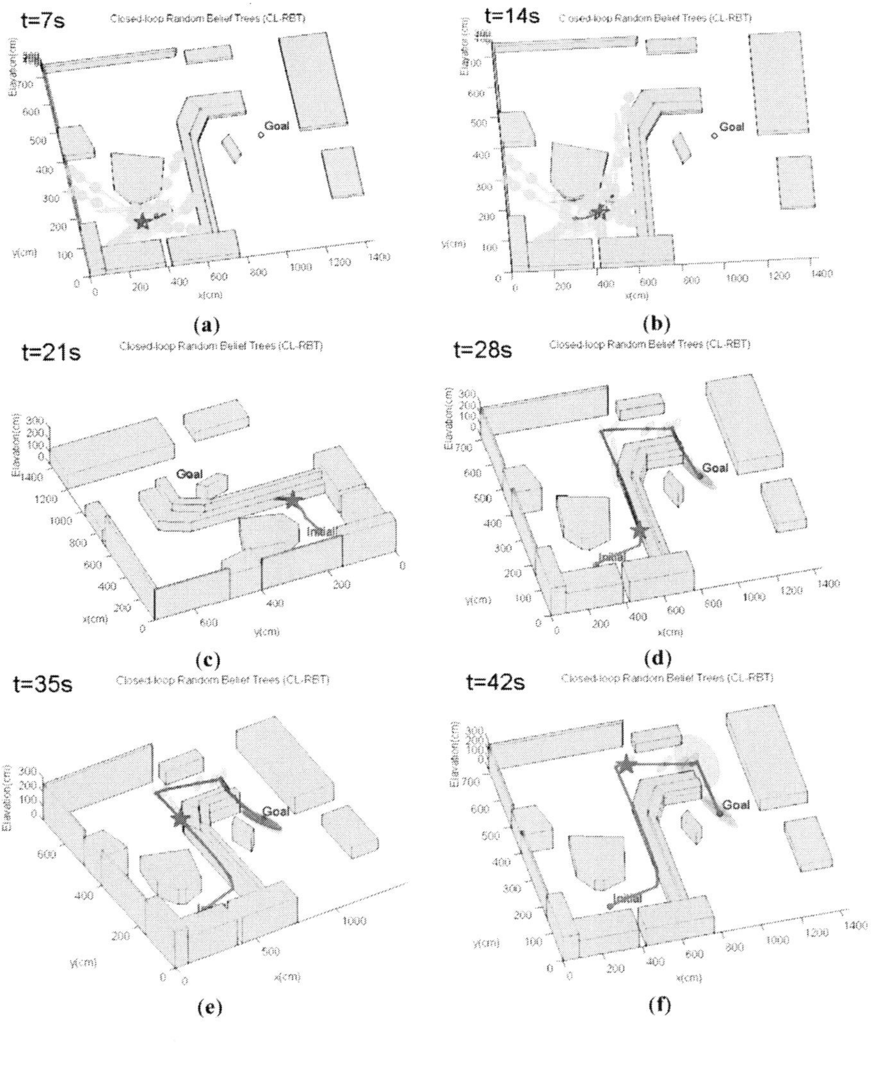

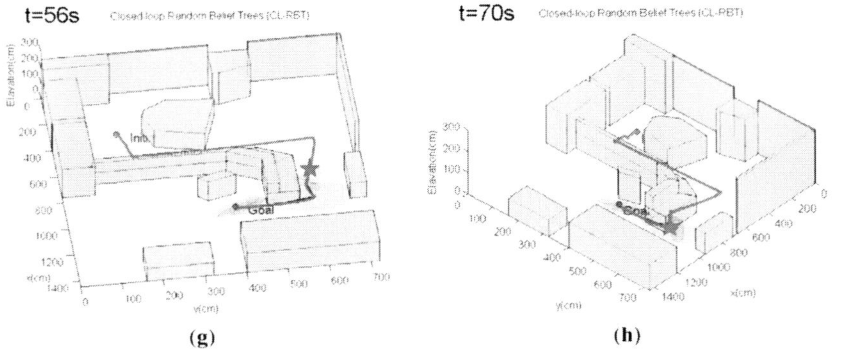

Figure 11. Example results of the trajectory generated by the CL-RBT for a quadrotor MAV navigating in a cluttered environment. The full legend for the symbols can be found in Figure 8. The current tree is set as invisible after (**b**) for clarity. (**a**) t = 7 s; (**b**) t = 14 s; (**c**) t = 21 s; (**d**) t = 28 s; (**e**) t = 35 s; (**f**) t = 42 s; (**g**) t = 56 s; (**h**) t = 70 s.

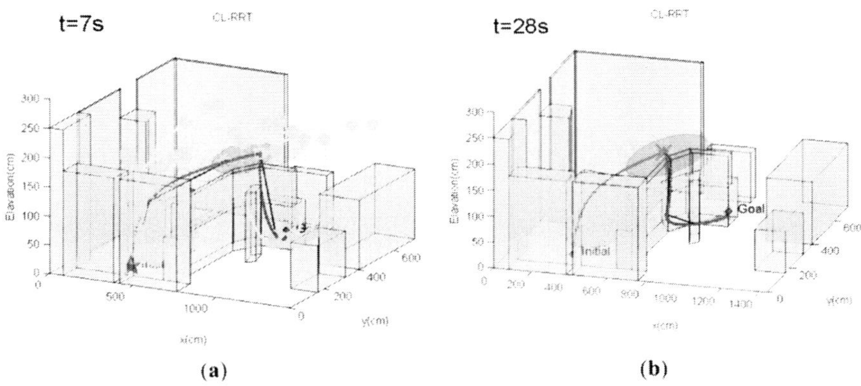

Figure 12. Example results of trajectory generated by the conventional CL-RRT for a quadrotor MAV navigating in simulation Scenario 1. The red cross in (**b**) denotes the position where the MAV's state estimation fails since the position uncertainty have exceeded the threshold. (**a**) t = 7 s; (**b**) t = 28 s.

6.5.3. Performance Comparison and Analysis

In order to illustrate the performance of the CL-RBT algorithm, we compare the performance statistics of the CL-RBT algorithm to that of the conventional CL-RRT by running simulation experiments using both algorithms on the same scenarios, as described in Sections 6.5.1 and 6.5.2 (note that we did not compare the CL-RBT to other conventional sampling-based path planners that operate on static graphs or trees, such as PRM, RRT or BRM, since they are not real-time algorithms in essence). For each simulation scenario, the experiment is repeated 30 times using each of the two algorithms, and the performance statistics is recorded and evaluated in terms of the paths'; cost (total length) and the trace of the MAV's expected covariance at the goal when following the generated paths, both averaged over 30 times.

Table 1 shows a comparison of the average path length and covariance statistics of CL-RBT and conventional CL-RRT. The CL-RBT outperforms conventional CL-RRT in terms of expected uncertainty at the goal in both simulation scenarios. Moreover, since these experiments are performed in simulation, the actual trajectories of the MAV are available, and the average mean error between the estimated MAV position and the actual position at the goal location are also computed and shown in Table 1 (note that the mean errors corresponding to the conventional CL-RRT are computed by assuming that the controller is able to execute the prescribed path and navigate the MAV to the goal location regardless of the estimation error, while in fact, the MAV controller would have failed halfway). These results indicate that the conventional CL-RRT results in large deviations from the true position, while the CL-RBT can effectively bound the estimation error in repeated trials. However, the resulting paths generated by CL-RRT are generally longer that those of conventional CL-RRT. This is because CL-RBT integrates the prediction of position uncertainty into the planning progress, which enabling the MAV to balance minimizing path length against reducing localization uncertainty, selecting longer paths to avoid regions that lead to high position estimation uncertainty when necessary.

Table 1. Performance comparisons of CL-RBT and CL-RRT, using 30 simulation runs.

Algorithms	Parameters	Scenario 1	Scenario 2
CL-RBT	Path length (m)	17.35	16.7
	Final covariance (cm^2)	**39.88**	**41.65**
	Final mean error (m)	1.07	0.82
CL-RRT	Path length (m)	10.74	12.78
	Final covariance (cm^2)	5.933	9.523
	Final mean error (m)	54.33	73.52

As discussed previously, the specification of weighting factors in the cost function for path selection Equation (19) is an important factor for implementations of CL-RBT. To evaluate the effect of the selection on weighting factors, we have conducted a number of simulations with different values of weighting factor ζ_3 in Scenario 1. Five different values of ζ_3 are evaluated in these simulations: $\zeta_3 = \{0, 1, 10, 50, 100\}$. For each of the five values, the simulation is repeated 10 times, and the path length and uncertainty cost (trace of the covariance) at the goal are recorded and averaged. Figure 13 illustrates the resulting paths' average length and uncertainty cost at the goal as a function of ζ_3. As can be seen from the figure, the uncertainty cost decreases as the weighting factor ζ_3 increases, whereas the path becomes comparatively longer. This is because reducing localization uncertainty becomes relatively more significant as ζ_3 increases, resulting in longer paths that take more measurements from the environment. These results demonstrate that the weighting factors of the cost function should be carefully selected depending on the specific task requirement and the characteristic of the environment.

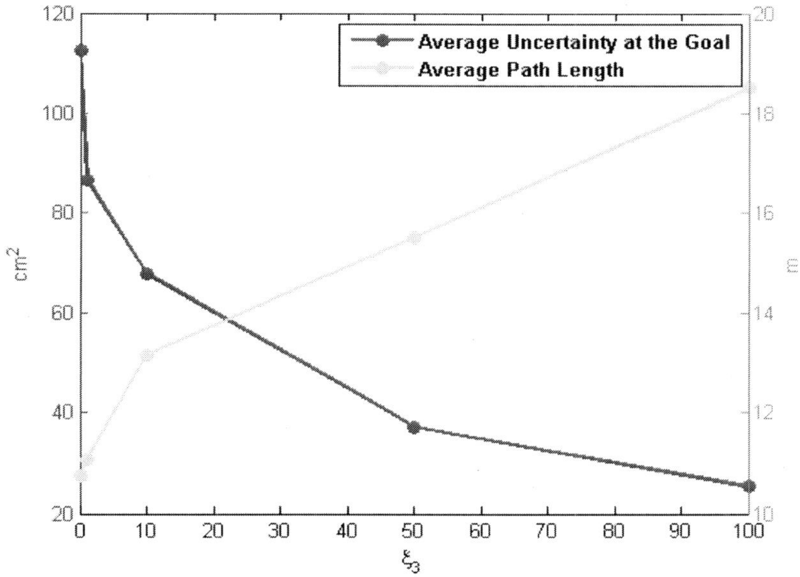

Figure 13. Comparison of average path length and average uncertainty cost at the goal. Data points correspond to different values of the uncertainty cost weighting factor (Equation (19)): ζ_3 = {0, 1, 10, 50, 100}.

7. CONCLUSIONS AND FUTURE WORK

This paper presents a real-time path planning approach in belief space (CL-RBT) for MAVs with complex dynamic constraints under state estimation uncertainty. The proposed CL-RBT approach is built upon the RRT motion planning framework. We made two primary contributions in this paper. First, the prediction of state uncertainty is incorporated in the path planning process, which allows the path planning strategy to find the path that reduces the state uncertainty and ensures state estimation confidence, using an efficient, linear update process of covariance in a factored form. Second, closed-loop prediction is used in the motion planning framework to generate dynamically feasible trajectories, enabling the motion planner to handle complex dynamic and environment constraints. Simulation results demonstrate that the motion planner can be implemented in real time, generating dynamically feasible paths that trade off minimizing path cost and reducing state uncertainty accuracy, while easily handling complex vehicle dynamics. It can also be concluded that CL-RBT has the potential to increase the autonomy of MAVs that operate in complex, GPS-denied environments.

While the utility and advances of the CL-RBT algorithm have been presented in this paper, there still remains significant future work on the proposed framework. Future work will focus on the more extensive analysis of the algorithm's performance, including the computational complexity and optimality of the generated paths. Further theoretical work is also necessary in the proof and analysis of the algorithm's completeness and convergence. Moreover, we are also planning to implement the CL-RBT on an actual quadrotor MAV platform that is currently under development. It would be possible to validate the performance of CL-RBT with realistic onboard sensors (laser rangefinders, RGB-D, cameras, *etc.*) through flight tests in actual indoor environments.

ACKNOWLEDGMENTS

This work is sponsored by the China High-Tech 863 Program, No. 2001AA415340 and No. 2007AA04Z1A6, the China Natural Science

Foundation, No. 61174168, the Aviation Science Foundation of China, No. 20100758002 and 20128058006.

AUTHOR CONTRIBUTIONS

D.L. proposed the CL-RBT approach, conducted the simulation experiments and wrote the paper. Q.L. and J.S. supervised this work, and N.C. provided advices on the derivation of the linear covariance propagation approach and the establishment of the EKF process model.

CONFLICTS OF INTEREST

The authors declare no conflicts of interest.

REFERENCES

1 Bachrach, A.; He, R.; Roy, N. Autonomous Flight in Unknown Indoor Environments. *Int. J. Micro Air Veh.* **2009**, *4*, 277–298.

2 Shaojie, S.; Michael, N.; Kumar, V. Autonomous Indoor 3D Exploration with a Micro-Aerial Vehicle. Proceedings of the 2012 IEEE International Conference on Robotics and Automation, Saint Paul, MN, USA, 14–18 May 2012; IEEE Press: New York, NY, USA, 2012.

3 Bry, A.; Bachrach, A.; Roy, N. State Estimation for Aggressive Flight in GPS-denied Environments Using Onboard Sensing. Proceedings of the 2012 IEEE International Conference on

Robotics and Automation, Saint Paul, MN, USA, 14–18 May 2012; IEEE Press: New York, NY, USA, 2012.

4 Abdelkrim, N.; Aouf, N.; Tsourdos, A.; White, B. Robust Nonlinear Filtering for INS/GPS UAV Localization. Proceedings of the Mediterranean Conference on Control and Automation, Ajaccio, France, 25–27 June 2008; IEEE Press: New York, NY, USA, 2008.

5 Michael, N.; Fink, J.; Kumar, V. Cooperative Manipulation and Transportation with Aerial Robots. *Auton. Robot.* **2010**,*30*, 73–86.

6 Chowdhary, G.; Sobers, D.M.; Pravitra, C.; Christmann, C.; Wu, A.; Hashimoto, H.; Ong, C.; Kalghatgi, R.; Johnson, E.N. Self-Contained Autonomous Indoor Flight with Ranging Sensor Navigation. *J. Guid. Control Dyn.* **2012**, *35*, 1843–1854.

7 Weiss, S.; Scaramuzza, D.; Siegwart, R. Monocular-SLAM based Navigation for Autonomous Micro Helicopters in GPS-denied Environments. *J. Field Robot.* **2011**, *28*, 854–874.

8 Huang, A.S.; Bachrach, A; Henry, P.; Krainin, M.; Maturana, D.; Fox, D.; Roy, N. Visual Odometry and Mapping for Autonomous Flight Using an Rgb-d Camera. Proceedings of the International Symposium on Robotic Research, Flagstaff, AZ, USA, 28 August–1 September 2011; pp. 1–16.

9 Frazzoli, E.; Dahleh, M.A.; Feron, E. Real-time motion planning for agile autonomous vehicles. *J. Guid. Control Dyn.***2002**, *25*, 116–129.

10 Samuel, P.; Roy, N. The Belief Roadmap: Efficient Planning in Belief Space by Factoring the Covariance. *Int. J. Robot. Res.* **2009**, *28*, 1448–1465.

11 Thrun, S.; Burgard, W.; Fox, D. Chap. 15; Partially Observable Markov Decision Process. In *Probabilistic Robotics*; MIT Press: Cambridge, MA, USA, 2005; pp. 513–542.

12 Kaelbling, L.; Littman, M.; Cassandra, A. Planning and Acting in Partially Observable Stochastic Domains. *Artif. Intell.* **1998**, *101*, 99–134.

13 Berg, V.D.; Patil, J.S.; Alterovitz, R. Motion Planning under Uncertainty Using Iterative Local Optimization in Belief Space. *Int. J. Robot. Res.* **2012**, *31*, 1263–1278.

14 Bai, H.; Hsu, D.; Lee, W.; Ngo, V. Monte Carlo Value Iteration for Continuous State POMDPs. Proceedings of the 9th International Workshop on the Algorithmic Foundations of Robotics, Singapore, Singapore, 13–15 December 2010; Springer: Berlin/Heidelberg, Germany, 2011.

15 Kavraki, L.E.; Svestka, P.; Latombe, J.C.; Overmars, M.H. Probabilistic Roadmaps for Path Planning in High-Dimensional Configuration Spaces. *IEEE Trans. Robot. Autom.* **1996**, *12*, 566–580.

16 LaValle, S.M. *Rapidly-Exploring Random Trees: A New Tool for Path Planning*; Technical Report; Iowa State University: Ames, IA, USA, 1998.

17 Kuwata, Y.; Teo, J.; Karaman, S.; Fiore, G.; Frazzoli, E.; How, J.P. Motion Planning in Complex Environments Using Closed-loop Prediction. Proceedings of the AIAA Guidance, Navigation, and Control Conference, 18–21 August 2008.

18 Kuwata, Y.; Teo, J.; Fiore, G.; Karaman, S.; Frazzoli, E.; How, J.P. Real-time motion planning with applications to autonomous urban driving. *IEEE Trans. Control Syst. Technol.* **2009**, *17*, 1105–1118.

19 Berg, V.D.; Abbeel, P.; Goldberg, K. LQG-MP: Optimized Path Planning for Robots with Motion Uncertainty and Imperfect State Information. *Int. J. Robot. Res.* **2011**, *30*, 895–913.

20 Bry, A.; Roy, N. Rapidly-exploring Random Belief Trees for Motion Planning under Uncertainty. Proceedings of the 2011 IEEE International Conference on Robotics and Automation, Shanghai, China, 9–13 May 2011; IEEE Press: New York, NY, USA, 2011.

21 He, R.; Prentice, S.; Roy, N. Planning in Information Space for a Quadrotor Helicopter in a Gps-denied Environment. Proceedings of the 2008 IEEE International Conference on Robotics and Automation, Pasadena, CA, USA, 19–23 May 2008; IEEE Press: New York, NY, USA, 2008.

22 Abraham, B.; Prentice, S.; He, R.; Henry, P.; Huang, A.S.; Krainin, M.; Maturana, D.; Fox, D.; Roy, N. Estimation, Planning, and Mapping for Autonomous Flight Using an RGB-D Camera in GPS-denied Environments. *Int. J. Robot. Res.* **2012**, *31*, 1320–1343.

23 Levine, D.; Luders, B.; How, J.P. Information-rich path planning with general constraints using rapidly-exploring random trees. Proceedings of the AIAA Infotech@ Aerospace Conference, Atlanta, GA, USA, 20–22 April 2012.

24 Kalman, E.R. A New Approach to Linear Filtering and Prediction Problems. *J. Fluid. Eng.* **1960**, *82*, 35–45

25 Smith, R.; Self, M.; Cheeseman, P. Estimating Uncertain Spatial Relationships in Robotics. In *Autonomous Robotic Vehicles*; Cox, I.J., Wilfong, G.T., Eds.; Springer: Orlando, FL, USA, 1990; pp. 167–193.

26 Qing, L.; Dachuan, L.; Qifan, W.; Liangwen, T.; Yan, H.; Yixuan, Z.; Nong, C. Autonomous navigation and environment modeling

for MAVs in 3-D enclosed industrial environments. *Comput. Ind.* **2013**, *64*, 1161–1177.

27 Park, S.; Deyst, J.; How, J.P. Performance and Lyapunov Stability of a Nonlinear Path-Following Guidance Method. *J. Guid. Control Dyn.* **2007**, *30*, 1718–1728.

28 Bachrach, A.; Prentice, S.; He, R.; Roy, N. RANGE—Robust autonomous navigation in GPS-denied environments. *J. Field Robot.* **2011**, *28*, 644–666.

29 Bachrach, A. Autonomous Flight in Unstructured and Unknown Indoor Environments. Master's Thesis, Massachusetts Institute of Technology, Cambridge, MA, USA, September 2009.

CHAPTER 8

DESIGN OF A STABILITY AUGMENTATION SYSTEM FOR AN UNMANNED HELICOPTER BASED ON ADAPTIVE CONTROL TECHNIQUES

Shouzhao Sheng * and Chenwu Sun

College of Automation Engineering, Nanjing University of Aeronautics and Astronautics, 29 YuDao St., Nanjing 210016, China

ABSTRACT

The task of control of unmanned helicopters is rather complicated in the presence of parametric uncertainties and measurement noises. This paper presents an adaptive model feedback control algorithm for an unmanned helicopter stability augmentation system. The proposed algorithm can achieve a guaranteed model reference tracking performance and speed up the convergence rates of adjustable parameters, even when the plant parameters vary rapidly. Moreover, the model feedback strategy in the algorithm further contributes to the improvement in the control quality of the stability augmentation system in the case of low signal to noise ratios, mainly because the model feedback path is noise free. The effectiveness and superiority of the proposed algorithm are demonstrated through a series of tests.

KEYWORDS

adaptive control; model feedback; stability augmentation; measurement noise; unmanned helicopter

1. INTRODUCTION

It is essential that the flight control system of an unmanned helicopter (UH) should be endowed with well-suited automatic capabilities to carry out flight missions. However, the flight performance of an UH is intimately dependent on the stability and control characteristics of the UH [1,2]. Unlike some mechanical systems with desirable structural properties, UH is normally an inherently unstable system without stability augmentation control strategy. Furthermore, some of the aerodynamic parameters vary with flight environments or system conditions, with the result that it is very difficult to design the stability augmentation system for an UH using the conventional control methods [3].

UH is also a complicated nonlinear dynamic system described by nonlinear differential equations. However, for simplicity in the design of controllers for an UH, the linearized models are normally derived from nonlinear differential equations if the UH operates around an operating point. Many linear control techniques for application to UH flight control systems have been proposed in literature, among which single-input, single-output (SISO) feedback control methods are by far the most common choices with few dependencies on dynamic models. In [4], a SISO PD control law is adopted and further optimized for both hovering and forward flight of the CMU-R50 UH. In [5], a SISO PID control law is implemented for automatic hovering of the Ursa Major 3 UH. The above SISO methods have the advantages of conceptual and computational simplicity. However they may decrease the stability and control qualities of UHs without considering parametric uncertainties and cross-couplings among axes.

Therefore, in order to improve the flight performance, a lot of research effort has been devoted to the design of advanced stability augmentation systems. Previous research reported in the literature includes gain scheduling [6], linear-quadratic regulation (LQR) or linear-quadratic Gaussian (LQG) approach [7], decentralized decoupled model predictive approach [8] and intelligent control methods like neural network [9] and fuzzy logic approach [10], *etc.* Several flight control systems using H_∞ control methods, which can provide the robust stability and performance for the systems subject to uncertainties and disturbances, have been designed for mini UHs. In [11], a H_∞ loop shaping technique is utilized for the stability augmentation system of Bell-205 helicopter. Mixed-norm optimization and weighted H_∞ mixed sensitivity optimization methods are respectively designed to improve the stability and maneuverability characteristics of UHs [12,13]. Although these methods have achieved acceptable flight performance, they still rely heavily on the plant model. More importantly, the above-mentioned methods fail to consider the adverse effects of parametric uncertainties and measurement noises on the flying qualities. Some existing adaptive techniques can accommodate the parametric uncertainties more effectively without considering measurement noises [14,15,16,17,18]. A novel modified model reference adaptive control (MRAC) strategy is developed with added noise [19]. However, this method only reduces the noise disturbance using a low-pass filter. Usually in practical use, those existing adaptive control methods can hardly minimize the adverse effect of measurement noises on the flying qualities. The NASA Marshall Space Flight Center has developed an adaptive augmenting control (AAC) algorithm for launch vehicles by adapting a well tuned classical control algorithm to unexpected environments or variations in vehicle dynamics. The AAC algorithm has been successfully tested in a relevant environment. However it needs to be further evaluated by the flight tests [20].

The essential parameter regulation schemes can reduce the complexity of a high performance control system design problem in the presence of parametric uncertainties and measurement noises. On this basis, this study aims to develop an adaptive model feedback control algorithm for a prototype unmanned helicopter stability augmentation system. The proposed adaptive algorithm can achieve a guaranteed model reference tracking performance and speed up the convergence rates of adjustable parameters, even when the plant parameters vary rapidly. Moreover, the model feedback control strategy in the algorithm can further improve the control quality of the stability augmentation system because the model feedback path is noise free. The experimental setups and the actual flight test results using the proposed algorithm are shown and the results are discussed.

This paper describes the control problem for the prototype UH stability augmentation system in Section 2. The adaptive model feedback control algorithm for the prototype UH stability augmentation system is presented in Section 3, and Section 4 provides the improvement of the algorithm. The flight test results are shown in Section 5. Finally, conclusions are drawn in Section 6.

2. THE CONTROL PROBLEM FOR THE PROTOTYPE UH STABILITY AUGMENTATION SYSTEM

The prototype UH, with net weight 180 Kg and height 1.9 m, is a vertical takeoff and landing aircraft which includes two coaxial rotors and a fuselage with toroidal portion, as shown in Figure 1. A duct is formed through the fuselage. A propeller assembly is mounted to the top portion of the fuselage with a main rotor, 4.4 m in diameter, above the fuselage. A ducted rotor assembly in fuselage is used to compensate the propeller antitorque as well as providing some fraction of lift. The coaxial rotors, main and ducted, rotate at 800 rpm in the opposite directions with the main rotor providing about 70% of lift, drag, and pitch and roll movements of UH and the ducted rotor providing about 30% of lift and yaw movement.

Figure 1. The prototype unmanned helicopter.

The stability augmentation system design for the prototype UH is a challenging task because the UH dynamics are highly nonlinear and subject to parametric uncertainties. In addition, the plant parameters change with the flight environments (e.g., the aerodynamic constants) or the system conditions (e.g., lift curve slopes). The adaptive control strategy is adopted to solve the above problems in this study.

By assuming that the fuselage is a rigid body and the main rotor speed is constant, we can describe the nonlinear kinematic equations associated with the six degrees of freedom (6-DOF) as an equivalent block in the following manner:

$$\dot{\bar{X}} = f(\bar{X},\bar{U},\Theta) \tag{1}$$

where $\bar{X}=[\bar{u},\bar{v},\bar{w},\bar{\theta},\bar{\phi},\bar{\psi},\bar{q},\bar{p},\bar{r}]^{\mathrm{T}}$ and $\bar{U}=[\bar{\delta}_e,\bar{\delta}_a,\bar{\delta}_r,\bar{\delta}_c]^{\mathrm{T}}$ represent the state vector and the control vector, respectively; Θ represents the unsteady aerodynamic parameters set, which is difficult and expensive to measure and is, therefore, not available in most cases. Of the state variables in $\bar{X},\bar{U}$, here $\bar{u},\bar{v},\bar{w}$ are forward velocity, lateral velocity and vertical velocity, respectively; $\bar{\theta},\bar{\phi},\bar{\psi}$ are pitch angle, roll angle and yaw angle, respectively; $\bar{q},\bar{p},\bar{r}$ are pitch rate, roll rate and yaw rate, respectively; $\bar{\delta}_e,\bar{\delta}_a,\bar{\delta}_r,\bar{\delta}_c$ are pitch cyclic, roll cyclic, ducted rotor collective and main rotor collective, respectively.

One common method to solve the control problem for the nonlinear system in Equation (1) is through linearization. The linearized model about a trim condition of the nonlinear dynamics is then represented as

$$\dot{\boldsymbol{X}} = \boldsymbol{AX} + \boldsymbol{BU} \tag{2}$$

where $\boldsymbol{X} = \tilde{\boldsymbol{X}} - \boldsymbol{X}_e = [u, v, w, \theta, \phi, \psi, q, p, r]^{\mathrm{T}}$ and $\boldsymbol{U} = \tilde{\boldsymbol{U}} - \boldsymbol{U}_e = [\delta_e, \delta_a, \delta_r, \delta_c]^{\mathrm{T}}$ are the increments at a specified trim condition; A and B are the system matrix and the control-input matrix, respectively. Accordingly, Xe and Ue are respectively the trim state and the trim input with respect to hovering, lifting or forward flight, which must satisfy the equation

$$f(\boldsymbol{X}_e, \boldsymbol{U}_e, \Theta) = 0 \tag{3}$$

Note that Xe and Ue, obtained by solving Equation (3), are unknown as well. Therefore, we define

$$\boldsymbol{X}_0 = \boldsymbol{X}_e + \boldsymbol{E}_x, \boldsymbol{U}_0 = \boldsymbol{U}_e + \boldsymbol{E}_u \tag{4}$$

X0 and U0 can also be regarded as the estimates of Xe and Ue, respectively. Equation (2) can then be rewritten as:

$$\dot{\boldsymbol{X}} = \boldsymbol{AX} + \boldsymbol{BU} + \boldsymbol{E} \tag{5}$$

where $\boldsymbol{X} = \tilde{\boldsymbol{X}} - \boldsymbol{X}_0, \boldsymbol{U} = \tilde{\boldsymbol{U}} - \boldsymbol{U}_0$ and $\boldsymbol{E} = \boldsymbol{AE}_x + \boldsymbol{BE}_u$. $\boldsymbol{E}$ is equivalent to an unknown input disturbance at the trim condition.

To avoid repetition, the present study is only focused on the stability augmentation system of the longitudinal axis to demonstrate the proposed adaptive algorithm, which, without loss of generality, can apply to other axes. Thus, referring to Equation (5), the linearized model of the longitudinal axis can be denoted by:

$$\dot{q} = M_u^{\dot{q}} u + M_v^{\dot{q}} v + M_w^{\dot{q}} w + M_\theta^{\dot{q}} \theta + M_\phi^{\dot{q}} \phi + M_\psi^{\dot{q}} \psi + M_q^{\dot{q}} q + M_p^{\dot{q}} p + M_r^{\dot{q}} r + M_{\delta_e}^{\dot{q}} \delta_e + M_{\delta_a}^{\dot{q}} \delta_a + M_{\delta_r}^{\dot{q}} \delta_r + M_{\delta_c}^{\dot{q}} \delta_c + M_t^{\dot{q}}$$

(6)

where $M_u^{\dot{q}}, M_v^{\dot{q}}, M_w^{\dot{q}}, M_\theta^{\dot{q}}, M_\phi^{\dot{q}}, M_\psi^{\dot{q}}, M_q^{\dot{q}}, M_p^{\dot{q}}, M_r^{\dot{q}}, M_{\delta_e}^{\dot{q}}, M_{\delta_a}^{\dot{q}}, M_{\delta_r}^{\dot{q}}$ and $M_{\delta_c}^{\dot{q}}$ represent the unknown time-varying aerodynamic parameters; $M_t^{\dot{q}}$ is the corresponding component of the vector E.

Given the corresponding wind tunnel test data, the effect of θ,ϕ,ψ and δr on q is negligible and can therefore be ignored, the linearized model of longitudinal axis can then be simplified as follows:

$$\dot{q} = M_u^{\dot{q}} u + M_v^{\dot{q}} v + M_w^{\dot{q}} w + M_q^{\dot{q}} q + M_p^{\dot{q}} p + M_r^{\dot{q}} r + M_{\delta_e}^{\dot{q}} \delta_e + M_{\delta_a}^{\dot{q}} \delta_a + M_{\delta_c}^{\dot{q}} \delta_c + M_t^{\dot{q}}$$

(7)

Overall, the complexity of the stability augmentation system design originates mainly from the parametric uncertainties together with measurement noises.

3. DESIGN OF THE STABILITY AUGMENTATION SYSTEM BASED ON AN ADAPTIVE MODEL FEEDBACK CONTROL ALGORITHM

3.1. Theorem 1

Consider the plant of the form of Equation (7), with $M_q^{\dot{q}} < 0$ and $M_{\delta_e}^{\dot{q}} > 0$. Assume that

$$\delta_e = k_u^{\dot{q}} u + k_v^{\dot{q}} v + k_w^{\dot{q}} w + k_p^{\dot{q}} p + k_r^{\dot{q}} r + k_{\delta_a}^{\dot{q}} \delta_a + k_{\delta_c}^{\dot{q}} \delta_c + k_t^{\dot{q}} + k_{q_m}^{\dot{q}} q_m + k_{\delta_{em}}^{\dot{q}} \delta_{em} \quad (8)$$

Where

$$k_{q_m}^{\dot{q}} = \left(k_{q_m}^{\dot{q}_m} - k_q^{\dot{q}} \right) / k_{\delta_e}^{\dot{q}}, \quad k_{\delta_{em}}^{\dot{q}} = k_{\delta_{em}}^{\dot{q}_m} / k_{\delta_e}^{\dot{q}} \quad (9)$$

and where $k_u^{\dot{q}}, k_v^{\dot{q}}, k_w^{\dot{q}}, k_q^{\dot{q}}, k_p^{\dot{q}}, k_r^{\dot{q}}, k_{\delta_e}^{\dot{q}}, k_{\delta_a}^{\dot{q}}, k_{\delta_c}^{\dot{q}}$ and $k_t^{\dot{q}}$ are adjustable parameters; qm is the output of the ideal decoupled model given by:

$$\dot{q}_m = k_{q_m}^{\dot{q}_m} q_m + k_{\delta_{em}}^{\dot{q}_m} \delta_{em} \quad (10)$$

$k_{q_m}^{\dot{q}_m}$ and $k_{\delta_{em}}^{\dot{q}_m}$ denote the model parameters determined according to ADS-33, δem the manipulated input. Thus, qasymptotically converge to qm as the adaptive laws are given by:

$$\begin{cases} \dot{k}_q^{\dot{q}} = -\rho_q(\kappa+e)q \text{ and } k_q^{\dot{q}} < 0, \\ \dot{k}_{\delta_e}^{\dot{q}} = -\rho_{\delta_e}(\kappa+e)\delta_{ei}, \\ \dot{k}_\chi^{\dot{q}} = \rho_\chi(\kappa+e)\chi, \\ \dot{k}_t^{\dot{q}} = \rho_t(\kappa+e), \end{cases} \tag{11}$$

where ρq, ρδe, ρχ and ρt are greater than zero; χ represents u,v,w,p,r,δa and δc, respectively; the tracking error is defined as:

$$e = q_m - q \tag{12}$$

the time-varying parameter κ should satisfy the following constraint:

$$\operatorname{sgn}(\kappa) = \operatorname{sgn}\left(\dot{e} - k_q^{\dot{q}} e\right) \tag{13}$$

the generalized input is denoted by:

$$\delta_{ei} = k_{q_m}^{\dot{q}} q_m + k_{\delta_{em}}^{\dot{q}} \delta_{em} \tag{14}$$

3.2. Proof

Substituting Equation (14) into Equation (8) yields

$$\delta_e = k_u^{\dot{q}} u + k_v^{\dot{q}} v + k_w^{\dot{q}} w + k_p^{\dot{q}} p + k_r^{\dot{q}} r + k_{\delta_a}^{\dot{q}} \delta_a + k_{\delta_c}^{\dot{q}} \delta_c + k_t^{\dot{q}} + \delta_{ei} \tag{15}$$

Again the generalized plant shown in Equation (16) is derived by substituting Equation (15) into Equation (7).

$$\dot{q} = M_q^{\dot{q}} q + M_{\delta_e}^{\dot{q}} \delta_{ei} + \hat{M}_u^{\dot{q}} u + \hat{M}_v^{\dot{q}} v + \hat{M}_w^{\dot{q}} w + \hat{M}_p^{\dot{q}} p + \hat{M}_r^{\dot{q}} r + \hat{M}_{\delta_a}^{\dot{q}} \delta_a + \hat{M}_{\delta_c}^{\dot{q}} \delta_c + \hat{M}_t^{\dot{q}} \tag{16}$$

where $\hat{M}_\bullet^{\dot{q}} = M_\bullet^{\dot{q}} + M_{\delta_e}^{\dot{q}} k_\bullet^{\dot{q}}$.

Similarly, we have

$$\dot{q}_m = k_q^{\dot{q}} q_m + k_{\delta_e}^{\dot{q}} \delta_{ei} \tag{17}$$

Then, substituting Equations (16) and (17) into Equation (12) yields:

$$\dot{e} = k_q^{\dot{q}} e + b_q q + b_{\delta_{ei}} \delta_{ei} + b_u u + b_v v + b_w w + b_p p + b_r r + b_{\delta_a} \delta_a + b_{\delta_c} \delta_c + b_t \tag{18}$$

where

$$\begin{cases} b_q = k_q^{\dot{q}} - M_q^{\dot{q}}, \\ b_{\delta_{ei}} = k_{\delta_e}^{\dot{q}} - M_{\delta_e}^{\dot{q}}, \\ b_\bullet = -M_{\delta_e}^{\dot{q}} k_\bullet^{\dot{q}} - M_\bullet^{\dot{q}}, \end{cases} \tag{19}$$

Furthermore, the adaptive laws are chosen so that certain stability conditions based on Lyapunov theory are satisfied. Consider the Lyapunov function candidate

$$V = e^2 + \lambda_q b_q^2 + \lambda_{\delta_{ei}} b_{\delta_{ei}}^2 + \lambda_u b_u^2 + \lambda_v b_v^2 + \lambda_w b_w^2 + \lambda_p b_p^2 + \lambda_r b_r^2 + \lambda_{\delta_a} b_{\delta_a}^2 + \lambda_{\delta_c} b_{\delta_c}^2 + \lambda_t b_t^2 \tag{20}$$

where $\lambda_q, \lambda_{\delta_{ei}}, \lambda_u, \lambda_v, \lambda_w, \lambda_p, \lambda_r, \lambda_{\delta_a}, \lambda_{\delta_c}$ and λ_t are greater than zero. We can therefore conclude that V˙ is negative definite as

$$\begin{cases} k_q^{\dot{q}} < 0, \\ \lambda_\chi \dot{b}_\chi = -\kappa\chi - e\chi, \\ \lambda_t \dot{b}_t = -\kappa - e, \end{cases} \tag{21}$$

in which χ represents $q, \delta_{ei}, u, v, w, p, r, \delta_a$ and δ_c, respectively. From Equations (19) and (21), we can easily deduce the adaptive laws given by Equation (11).

3.3. Remarks

i. The direct addition of an external feedback term like a·q to δe is necessary without any modification of the adaptive laws if $M_q^{\dot{q}}$ cannot satisfy $M_q^{\dot{q}} < 0$ at a certain flight condition, where a is used to guarantee the stability of the plant, that is, $M_q^q + M_{\delta_e}^q a < 0$ in this case.

ii. In consideration of the adverse effect of measurement noises, the variance of the noise component of δe can be reduced while using the proposed adaptive algorithm, mainly because the model feedback path is noise free, which can result in the improvement in the control quality of the stability augmentation system, especially in the case of low signal to noise ratios (SNR).

iii. Compared to the model reference adaptive control (MRAC) method, the additional time-varying parameter κ in the proposed algorithm can contribute to the convergence rates of adjustable parameters when the plant parameters vary rapidly.

4. THE IMPROVEMENT AND APPLICATION OF THE ADAPTIVE LAWS

The adaptive laws are difficult to apply in practice because q˙ is non-measurable in most cases. According to linear system theory, Equation (18) can be rewritten as follows:

$$\dot{e}_f = k_q^q e_f + b_q q_f + b_{\delta_{ei}} \delta_{eif} + b_u u_f + b_v v_f + b_w w_f + b_p p_f + b_r r_f + b_{\delta_a} \delta_{af} + b_{\delta_c} \delta_{cf} + b_t t_f \tag{22}$$

where $e_f, q_f, \delta_{eif}, u_f, v_f, w_f, p_f, r_f, \delta_{af}, \delta_{cf}$ and t_f are respectively served as the outputs of an arbitrarily chosen filter Gf(s) in response to $e, q, \delta_{ei}, u, v, w, p, r, \delta_a, \delta_c$ and $1(t)$. The adaptive laws, derived by using the similar Lyapunov function described in Equation (20), are rewritten as:

$$\begin{cases} \dot{k}_q^q = -\rho_q\left(\kappa_f + e_f\right)q_f \ \text{ and } \ k_q^q < 0, \\ \dot{k}_{\delta_e}^q = -\rho_{\delta_e}\left(\kappa_f + e_f\right)\delta_{eif}, \\ \dot{k}_\chi^q = \rho_\chi\left(\kappa_f + e_f\right)\chi_f, \\ \dot{k}_t^q = \rho_t\left(\kappa_f + e_f\right), \end{cases} \tag{23}$$

where κf satisfies the following constraint:

$$\operatorname{sgn}\left(\kappa_f\right) = \operatorname{sgn}\left(\dot{e}_f - k_q^q e_f\right) \tag{24}$$

The improved adaptive laws are therefore subject to a certain form of the filter. Assume that the filter transfer function is given by:

$$G_f\left(s\right) = \frac{1}{Ts+1}, \tag{25}$$

where T is the time constant. In this case, Equation (24) can be written as:

$$\operatorname{sgn}\left(\kappa_f\right) = \operatorname{sgn}\left\{\frac{1}{T}\left(e - e_f\right) - k_q^q e_f\right\} \tag{26}$$

which indicates that κf is available. Figure 2 shows the schematic diagram of the adaptive stability augmentation system with model feedback. Note that the proper choice of κf can speed up the convergence of adjustable parameters.

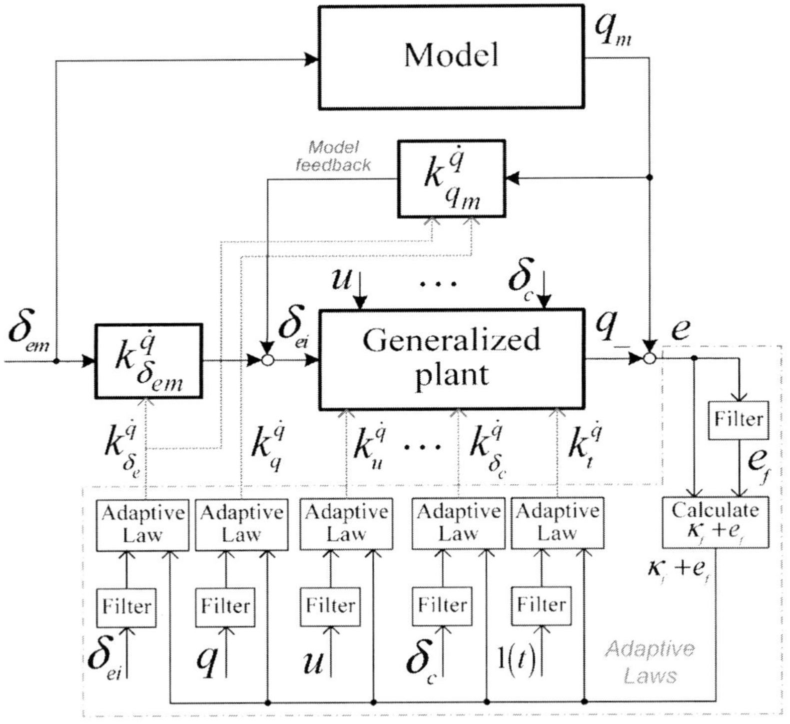

Figure 2. The schematic diagram of the adaptive stability augmentation system.

Note that $k_{q_m}^{\dot{q}}$ and $k_{\delta_{em}}^{\dot{q}}$ are completely dominant over other adjustable parameters in Equation (8), however the noise component of qm can be ignored on the basis of the assumption that the manipulated input δem is noise free. For δe, the adaptive model feedback strategy can therefore lead to a marked loss in noises.

5. FLIGHT TESTS

In this section, the performance of the proposed algorithm is demonstrated for the stability augmentation system of the longitudinal axis. All pre-chosen parameters are given as follows:

1) The model parameters: $k_{q_m}^{q_m} = -1.5$ and $k_{\delta_{em}}^{q_m} = 2.25$;
2) The initial values of adjustable parameters: $k_u^{\dot{q}}(0), k_v^{\dot{q}}(0), k_w^{\dot{q}}(0), k_q^{\dot{q}}(0), k_r^{\dot{q}}(0), k_{\delta_a}^{\dot{q}}(0), k_{\delta_c}^{\dot{q}}(0)$ and $k_t^{\dot{q}}(0)$ are assumed to equal zero, and $k_p^{\dot{q}}(0) = 0, k_{\delta_e}^{\dot{q}}(0) = 1.0$;
3) The time constant of the filter: T=0.02;
4) The parameters of the adaptive control laws: $\rho_\bullet = 50$; $\kappa_f = 2.0\left[\frac{1}{T}(e - e_f) - k_q^{\dot{q}} e_f\right]$; $a = -0.2$.

The proposed algorithm is first implemented in a numerical simulation based on the nonlinear helicopter model. It is then applied to the prototype UH to evaluate the flight performance of the stability augmentation system.

5.1. Task 1: Numerical Simulation

The numerical simulation test, shown in Figure 3, is conducted using the nonlinear dynamics model of the prototype unmanned helicopter, together with a simplified closed-form trim calculation. The manipulated input signal is generated from a joystick device. Various moderately aggressive maneuvers are conducted during the simulation to evaluate the performance of the proposed algorithm at different operate points of the flight envelope. A comparison is made between the output of the plant and that of the model, as shown in Figure 3a,b, which demonstrates that the tracking error can rapidly approach zero in 0.9 s and remain in a bounded range. We can therefore conclude that the proposed adaptive algorithm achieves a guaranteed model reference tracking performance and has a good convergence property. The tracking error remains in a bounded range even though the adjustable parameters vary rapidly during the test. However the conventional MRAC approach is not applicable to the case where the variations of the plant parameters are significant.

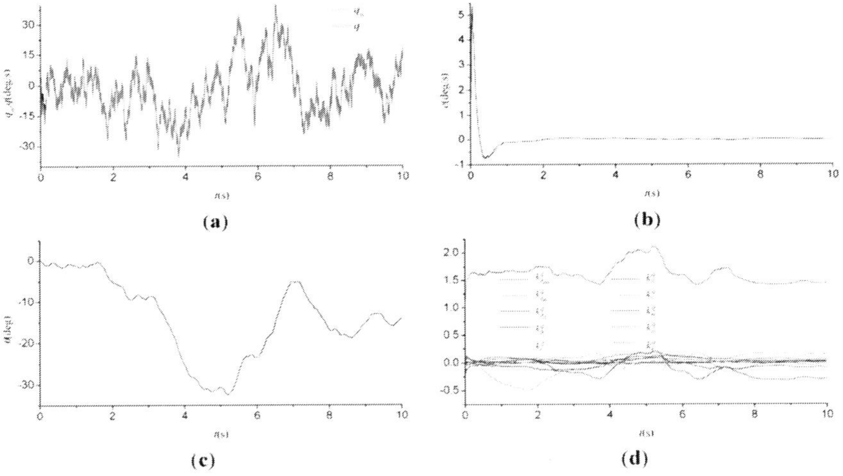

Figure 3. Numerical simulation results of the proposed algorithm. (**a**) qm and q; (**b**) e; (**c**) θ; (**d**) Adjustable parameters.

5.2. Task 2: Flight Tests

The flight tests are conducted to compare the performance of the proposed algorithm with that of the conventional PID while those two approaches applies to the prototype unmanned helicopter stability augmentation system, respectively. For a fair comparison, the attitude and position controllers remain unchanged. Our control software system, which runs on a DSP-based hardware platform with a control period of 10 ms, can provide reliable support for high precision timer and synchronization operations. The sensors (including an AHRS unit and a DGPS unit) are used to provide the information related to the flight status, such as the angles, angular rates, velocity and position. During flight, Wi-Fi and serial-based data links provide a link to the ground station computer that allows monitoring the real-time flight information and uploading remote control commands such as way points. Testing of the proposed adaptive algorithm begins with hover, followed by simple way-point navigation due to the limitations of the test conditions such as flight safety. In consideration of measurement noises of the sensors, the results from these flight tests are provided in Figure 4 and Figure 5.

The performance of the adaptive stability augmentation system is first evaluated at low speeds where a square pattern is flown, as shown in Figure 4. The noisy tracking error, shown in Figure 4b, is also within an acceptable range. Moreover, the model feedback strategy in the proposed algorithm contributes to the restricted variation of the control signal δe shown in Figure 4e, which explicitly improves the external command position tracking performance, and implicitly demonstrates the improvement in the control quality of the stability augmentation system.

By comparison with the above-mentioned results, the flight test results of the conventional PID controller implemented through the widely used Pixhawk autopilot module, shown in Figure 5, appear to deteriorate. For instance, the variation of the control signal δe, together with the external command position tracking errors, increases when the prototype UH is commanded to perform the same flight mission.

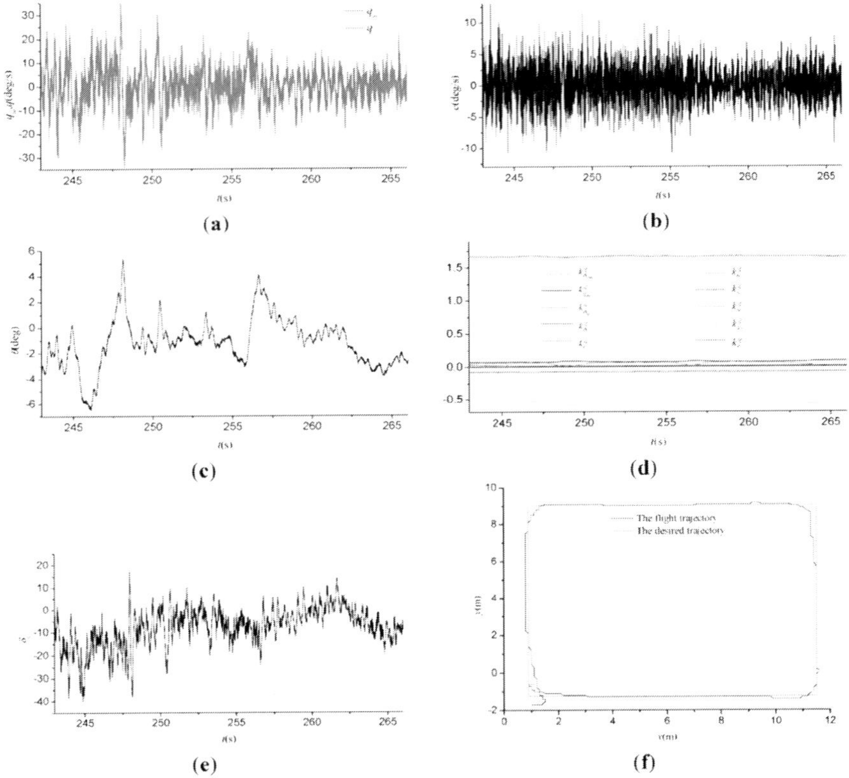

Figure 4. Flight test results of the proposed algorithm. (**a**) qm and q; (**b**) e; (**c**) θ; (**d**) Adjustable parameters; (**e**) δe; (**f**) The ground track view of trajectory.

6. CONCLUSIONS

This paper presents the adaptive model feedback control algorithm for the prototype UH stability augmentation system in the presence of parametric uncertainties and measurement noises. The proposed adaptive algorithm is able to achieve a guaranteed model reference tracking performance and speed up the convergence rates of adjustable parameters, even when the plant parameters vary rapidly. Moreover, the model feedback strategy in the proposed algorithm further contributes to the improvement in the control quality of the stability augmentation system in the case of low SNR. The flight test results have shown that the proposed algorithm can considerably improve the flying qualities of the prototype UH.

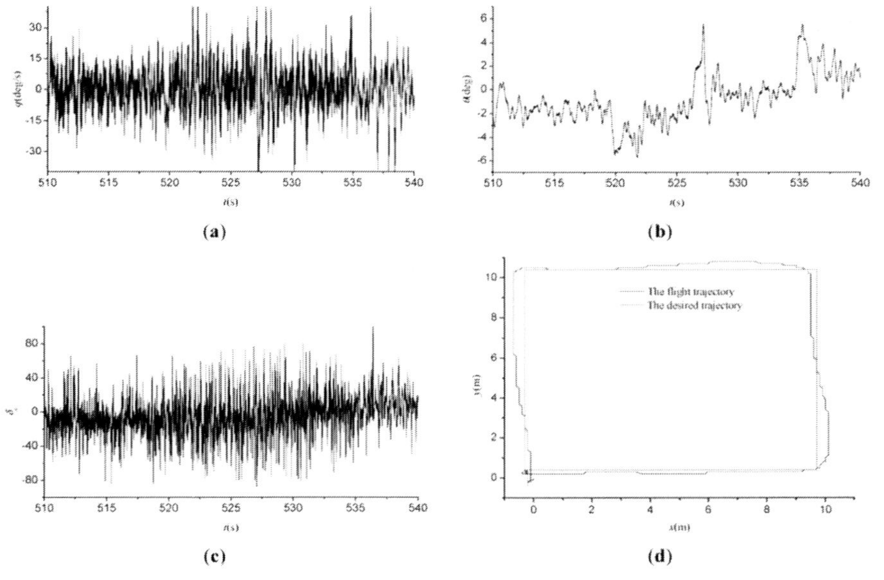

Figure 5. Flight test results of the PID controller. (**a**) q; (**b**) θ; (**c**) δe; (**d**) The ground track view of trajectory.

Based on the flight test results, we can conclude that the proposed adaptive model feedback control algorithm is able to adapt to rapidly changing flight conditions and effectively enhance the performance of the prototype UH stability augmentation system in the presence of parametric uncertainties and measurement noises, which results in the improvement in flying qualities. More significantly, most existing designs would require accurate models at each point, while the proposed design does not.

ACKNOWLEDGMENTS

This study was supported in part by National Natural Science Foundation of China (NSFC) (Under Grant No. 61374188), Aeronautical Science Foundation of China (Under Grant No. 2013ZC52033), Natural Science Foundation of Jiangsu Province of China (Under Grant No. BK20141412).

AUTHOR CONTRIBUTIONS

All authors discussed the contents of the manuscript. Shouzhao Sheng contributed to the research idea and the framework of this study. Chenwu Sun performed the experimental work.

CONFLICTS OF INTEREST

The authors declare no conflict of interest.

REFERENCES

1 Cai, G.; Chen, B.M.; Lee, T.H. *Unmanned Rotorcraft Systems*, 1st ed.; Springer: New York, NY, USA, 2011.

2 Ren, B.; Ge, S.S.; Chen, C.; Fua, C.H.; Lee, T.H. *Modeling, Control and Coordination of Helicopter Systems*, 1st ed.; Springer: New York, NY, USA, 2012.

3 Cook, M.V. *Flight Dynamics Principles: A Linear Systems Approach to Aircraft Stability and Control*, 3rd ed.; Butterworth-heinemann: Oxford, UK, 2012.

4 Mettler, B. *Identification Modeling and Characteristics of Miniature Rotorcraft*, 1st ed.; Springer: New York, NY, USA, 2011

5 Shim, D.H.; Kim, H.J.; Sastry, S. Control System Design for Rotorcraft-Based Unmanned Aerial Vehicles Using Time-Domain System Identification. In Control Applications, Proceedings of the 2000 IEEE International Conference on, Anchorage, AK, USA, 25–27 September 2000; pp. 808–813.

6 Takahashi, M.; Schulein, G.; Whalley, M. Flight Control Law Design and Development for an Autonomous Rotorcraft. In Proceedings of the 64th Annual Forum of the American Helicopter Society, Montreal, Quebec, Canada, 29 April–1 May 2008; pp. 1652–1671.

7 How, J.; Bethke, B.; Frank, A.; Dale, D.; Vian, J. Real time indoor autonomous vehicle test environment. *IEEE Control Syst. Mag.* **2008**, *28*, 51–64.

8 Shim, D.H.; Kim, H.J.; Sastry, S. Decentralized Nonlinear Model Predictive Control of Multiple Flying Robots. In Decision and Control, Proceedings of 42nd IEEE Conference on, Maui, HI, USA, December 2003; pp. 3621–3626.

9 Buskey, G.; Wyeth, G.; Roberts, J. Autonomous Helicopter Hover Using an Artificial Neural Network. In Proceedings of the IEEE Conference on Robotics and Automation, Seoul, Korea, 21–26 May 2001; pp. 1635–1640.

10 Garcia, R.; Valavanis, K. The implementation of an autonomous helicopter testbed. *J. Intell. Robot. Syst.* **2009**, *54*, 423–454.

11 Smerlas, A.J.; Walker, D.J.; Postlethwaite, I.; Strange, M.E.; Howitt, J.; Gubbels, A.W. Evaluating H_∞ controllers on the NRC Bell 205 fly-by-wire helicopter. *Control Eng. Pract.* **2001**, *9*, 1–10.

12 Walker, D.J. Multivariable control of the longitudinal and lateral dynamics of a fly-by-wire helicopter. *Control Eng. Pract.* **2003**, *11*, 781–795.

13 Gadewadikar, J.; Lewis, F.; Subbarao, K.; Chen, B.M. Structured H-infinity command and control-loop design for unmanned helicopters. *J. Guid. Control Dynam.* **2008**, *31*, 1093–1102.

14 Dytek, Z.T. Adaptive Control of Unmanned Aerial Systems. Ph.D. Thesis, Massachusetts Institute of Technology, Cambridge, MA, USA, September 2010.

15 Krupadanam, A.S.; Annaswamy, A.M.; Mangoubi, R.S. Multivariable adaptive control design with applications to autonomous helicopters. *J. Guid. Control Dynam.* **2002**, *25*, 843–851.

16 Liu, Y.; Tao, G. Modeling and model reference adaptive control of aircraft with asymmetric damage. *J. Guid. Control Dynam.* **2010**, *33*, 1500–1517.

17 Dauer, J.C.; Faulwasser, T.; Lorenz, S.; Findeisen, R. Optimization-Based Feed Forward Path Following for Model Reference Adaptive Control of an Unmanned Helicopter. In Proceedings of the AIAA Guidance, Navigation, and Control Conference, Boston, MA, USA, 19–22 August 2013.

18 Rusnak, I.; Weiss, H.; Barkana, I. Improving the performance of existing missile autopilot using simple adaptive control. *Int. J. Adapt Control Signal Process.* **2014**, *28*, 732–749.

19 Neild, S.A.; Yang, L.; Wagg, D.J. A modified model reference adaptive control approach for systems with noise or unmodelled dynamics. *Proc. Inst. Mech. Eng., Part I: J. Syst. Control Eng.* **2008**, *222*, 197–208.

20 VanZwieten, T.; Gilligan, E.; Wall, J.; Orr, J.; Miller, C.; Hanson, C. Adaptive Augmenting Control Flight Characterization Experiment on an F/A-18. In Proceedings of American Astronautical Society Guidance & Control Conference, Breckenridge, CO, USA, 31 January–5 February 2014; Volume 1, pp. 1–17.

INDEX